DemoGraphics

packaging

Published and distributed by
RotoVision SA
Route Suisse 9, CH-1295
Mies, Switzerland

RotoVision SA
Sales and Editorial Office
Sheridan House
114 Western Road
Hove, BN3 1DD, UK
Tel: +44 (0) 1273 72 72 68
Fax: +44 (0) 1273 72 72 69
E-mail: sales@rotovision.com
Web: www.rotovision.com

10 9 8 7 6 5 4 3 2 1

978-2-940361-71-7

Art Direction: Tony Seddon
Design and design palettes:
Absolute Zero°
Artwork: Spike
Illustrations: Ian Keltie
www.iankeltie.com

Reprographics in Singapore
by ProVision Pte. Ltd.
Tel: +65 6334 7720
Fax: +65 6334 7721

Printed in China by
Midas Printing International Ltd.

DemoGraphics

packaging

DESIGN SUCCESSFUL PACKAGING
FOR SPECIFIC CUSTOMER GROUPS

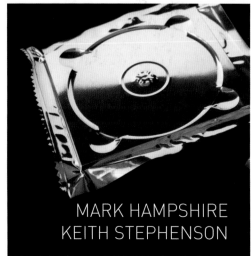

MARK HAMPSHIRE
KEITH STEPHENSON

CONTENTS_

packaging

INTRODUCTION BY MICHAEL PETERS, IDENTICA

A recent US survey found that at 4pm in the afternoon 60 percent of those questioned said they didn't know what they would be buying for dinner that evening, and would make their decision when they shopped. That's a staggering 60 percent of choices made at point of sale. If ever there was a case for the power of packaging design, this is it. And if a pack can trigger the decision to buy, it goes without saying that great packaging design can truly enhance or transform a product and a brand.

I have spent a good deal of my career expounding the virtues of good packaging design to brand owners, marketing directors, and entrepreneurs. My conviction is that great packaging is at the heart of the brand, reflecting its values and delivering the promise at the point of purchase. I also believe passionately in the power of innovation to differentiate and set new standards in the increasingly competitive marketplace. And, most pertinent to the topic covered by this book, I am convinced that focused customer segmentation is vital in creating effective packaging solutions.

So why is it so important to know your target audience? Though I am often asked about the merits of directing a brand at just a small part of the population, the profits to be made from correct targeting are legendary. To illustrate this point, let's take two examples of how to target different sectors with beauty products.

The first example is Penhaligon's. This brand is targeted at sophisticated, mature women who are serious about high quality, are quite traditional, and prepared to pay for the best. The design of the brand plays perfectly to this market by inventing a Victorian heritage, complete with royal crests, apothecary-inspired glass bottles, ribbons, and wax seals. Traditional, elegant, and expensive—exactly the same adjectives we would use to describe the target audience. A reflection, if you like, of the consumer's personality and values.

On the other side of the coin is Pout. Vivacious and a bit provocative, this brand is aimed at young professional women who value being up to date with the latest trends and care as much about the way their cosmetics are packaged as they do about the handbag they carry. Founded by three women who are themselves the epitome of the target audience, this is a brand with a strongly identifiable personality and packaging to match.

Two very different ways to sell beauty products by focusing on specific audiences. How these two brands' owners developed their insights into their target consumers is the subject of a deeper debate. In the case of Penhaligon's, the target consumer's desire for brands with heritage and authenticity are central to the product proposition and the packaging design. Pout's founders identified the need for the brand because there was nothing on the market for women like them—so the brand reflects their personal aspirations and desires.

Approached to create distinctive packaging for the Urban Garden Honey Co., we understood the target consumer to be someone who appreciates premium offerings with interesting provenance. Knowing that this consumer could be a man or a woman, the design needed to be appealing to both genders. In addition, we took into account the accepted visual codes of gourmet or premium packaging: understated black-and-white photography and bold, contemporary typography. What we arrived at tells the distinctive story of the honey's provenance. The pack designs highlight the unique urban personality of this quirky brand using photographs of the stone-carved flowers found in the city streets near the Urban Garden Honey Co. beehives. →

OPPOSITE:
The Urban Garden Honey Co. range
Design: Identica

→ However consumer insights are generated, the process of turning them into effective packaging design is the same. The need states of the target consumer are identified, the brand is built upon specific benefits pertinent to its target consumer, and the packaging design is created to reflect the consumer needs and brand benefits. Every successful brand does this in order to create distinctiveness and differentiation. Hence every brand has a defined personality and a distinctive packaging style to help target its consumer. Brands need to know their customers inside out.

Identica works extensively in the Russian market, often helping leading businesses optimize their brands in the face of competition from international imports. We have to be very sensitive to the Russian consumer's expectations, balancing national heritage with international quality cues. We created a strong new positioning, brand identity, and packaging for Flagman, the biggest selling standard vodka in Russia. Moving it away from its naval heritage was key to losing the old Soviet associations and the brand now appears contemporary, fresh, and proud, with great standout in an increasingly crowded market.

As designers, it is our job to turn the marketer's brief into effective communication solutions. That can sometimes seem daunting—especially when the brief comes packed with charts and statistics and peppered with the type of jargon that marketers like to use. "We've identified this type of housewife as a 'Faithful Dog' because she is devoted to her family's well-being." We've all been there. But simplicity is often the best approach, and by clearly visualizing your end user, simplifying all those purchasing statistics, need states, and brand benefits into key themes that will be relevant to the consumer, great design can be achieved.

Through packaging design, we attempt to express a wealth of values and benefits on the surface of a box, a bottle, or a can. I often encourage brand owners to think of it as the best-value advertising they can buy. It's there 365 days a year, in front of the target consumer in-store. And—as the statistic I mentioned earlier shows—that's the most important place because it's where the purchasing decision is most usually made. Packaging needs to woo its target consumer with great design, tactile materials, and exciting shapes. It needs to clearly differentiate the brand from its competitors. If it's done well, it benefits the bottom line in two ways, by boosting sales and adding value. Brand owners get a superb return on investment.

But ROI can mean more than just Return on Investment. I like to think of it as Return on Innovation. By developing innovative and unique pack design, brands can immediately generate differentiation and gain competitive advantage. We talk a lot about "creating standout" in this business; my frustration is with brand owners who say they want standout, yet don't have the strength of their convictions when presented with a piece of truly revolutionary design. Innovation means many things. It means using consumer insights more creatively, looking beyond the sector for influences that might change the consumer's attitude toward a product. It means being forward thinking about the use of consumer research.

We live in a world where consumer decisions are made on the run, on impulse, on the spur of the moment. And that means people's attitudes are changing rapidly. All businesses need to respond to their changing demands. As designers, we need to embrace change too. The best packaging designers aren't just people who organize the labels nicely. They're social observers, futurologists, marketers, designers, scientists—they're engineering the brand, not just the box.

Michael Peters, OBE, is Founder, CEO, and Chairman of Identica UK

OPPOSITE:
Flagman Vodka
Design: Identica

CHAPTER 1 _KIDS _M/F

It's no longer simply toys that are marketed to children. Research has identified that children play an active role in the decision-making process when it comes to purchasing a whole range of products for them—food, clothes, even technology. Marketers have realized the potential of pester power. Given the complex negotiating process that occurs between parent and child at the point of purchase, it's no surprise that packaging has to tick two boxes: fun for kids and reassuring for parents. Perhaps with this balance in mind, the gaudy approach to visual styling is slowly giving way to a more design-led and informative packaging style, inspiring bright kids everywhere.

CASE STUDY _SAINSBURY'S KIDS

"These days, children absorb
a wealth of information about
health and nutrition from school
and other sources, so they are
active participants in the
purchasing decision."

KATE BRADFORD
MANAGING DIRECTOR, PARKER WILLIAMS

When creating the packaging for Sainsbury's range of healthy food for kids, design consultancy Parker Williams had two audiences to consider. The packaging needed to appeal to the purchaser parent and the consumer child alike. "But it's no longer simply a case of healthy for mom and fun for kids," explains Managing Director, Kate Bradford. "Children are demanding consumers. The product has to taste good and have a degree of playground credibility too."

Kate is a firm believer that the most useful consumer research is done at the pre-design stage. "That way, it's not simply about choosing one design route over another, but about uncovering useful insights that can help shape the creative brief." In this case, kids in focus groups were particularly drawn to images of children having fun with food. "One image in particular—of a child holding a watermelon slice as a smile—really resonated with children and formed the basis for the creative route developed."

The Firefly range was created in response to research showing that kids wanted cellphones, but parents were reluctant to purchase them due to concerns over lack of control. Designed to keep 8–12 year olds connected to the people most important to them, the phone has speed dials for mom and dad. Colorful, but not overly childish, its packaging is the interchangeable covers and see-through accessories packs with bright and playful branding by Factor Design.

With his trademark ears, Mickey Mouse is one of the world's most recognizable icons. Factor Design's work for Disney's consumer electronics division puts Mickey center stage, giving the TV packaging prominence on the shelves of Target, Best Buy, and Toys R Us. It's testament to the brand's strength that it can transfer so effortlessly from electrical goods to toys—especially in the hands of Parker Williams, whose Disney Magic Artist Dough packaging brings the magic to life on pots and packs.

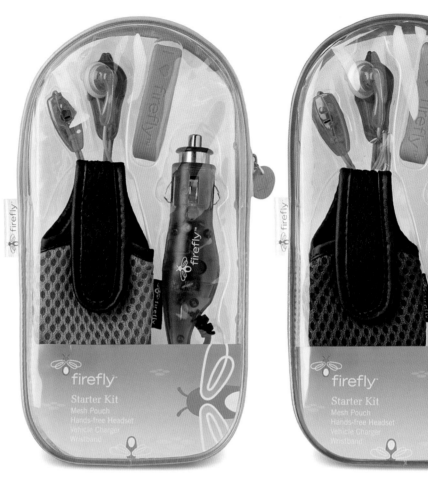

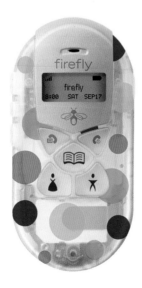

THIS PAGE:
Firefly cellphones accessories
and handset packaging
Design: Firefly
Branding: Factor Design

OPPOSITE TOP:
Disney's Mickey Mouse TV packaging
Design: Factor Design

**OPPOSITE MIDDLE
AND BOTTOM:**
Disney's Magic Artist Dough
Design: Parker Williams

KIDS_M/F
BRIGHT KIDS

CASE STUDY _EARLY LEARNING CENTRE

"Packaging is certainly one of
the key reasons behind our improved
sales. It's all part of the snowball
we have created with new stores and
new products. Compared to where we
were two years ago, customers
and staff love it."

TIM PATTEN

MARKETING DIRECTOR, EARLY LEARNING CENTRE

The children's retailer, Early Learning Centre (ELC) has recently conducted a review and revamp of its stores, products, and packaging under the guidance of Marketing Director, Tim Patten. He points out that understanding the needs of its core customer has been key to success. "Our customers are women aged 30-plus—moms of children aged 0–6 years—with an interest in being involved in helping their children learn as they play."

ELC toys are designed to help children explore the boundaries of their imaginations and creativity, to make learning fun, and help children be all they can be. This philosophy is at the heart of the packaging design. Tim explains: "because parents who shop with us have an interest in helping their children learn as they play, we have incorporated icons on the packaging to show what each toy can help develop—such as social skills, imagination, and so on."

The branding and design consultancy, Parker Williams created packaging style guides and helped to introduce a color-coding system to clarify the categories according to children's learning and playing need states.

The 11 categories include "art and creativity," "let's pretend," "action and adventure," "learning is fun," and "puzzle it out." In addition, there are sub-brands such as Happyland, Planet Protectors, Soft Stuff, and the ever-popular line of Early Learning Classics.

"Another key innovation," says Tim, "has been to incorporate playing tips to help moms with ideas on how to play with their children."

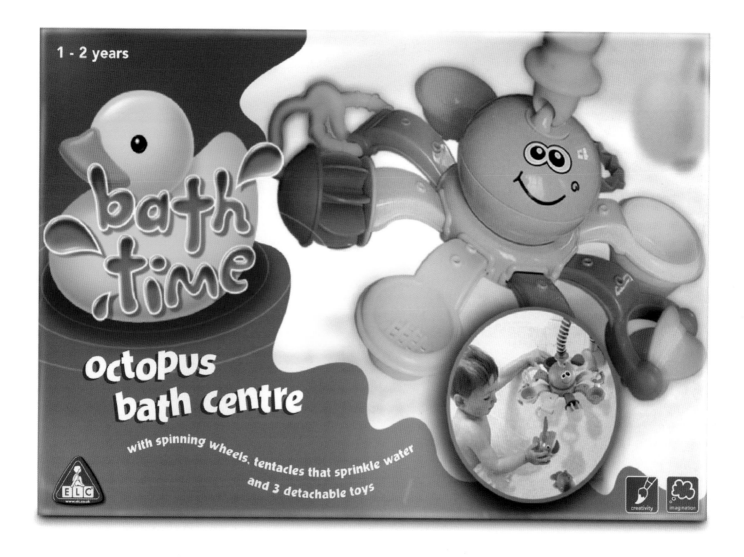

FOCUS _EARLY LEARNING CENTRE

The friendly rounded-edged highlight box is used across all categories of toy to draw attention to specific product features and benefits.

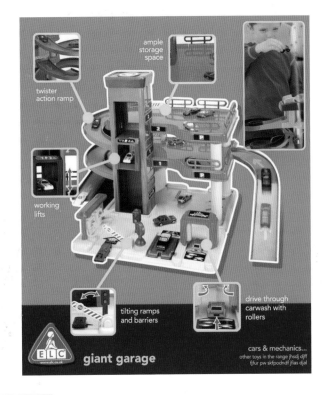

ample storage space

twister action ramp

working lifts

tilting ramps and barriers

drive through carwash with rollers

cars & mechanics...
other toys in the range jhsdj djff fjfur pw skfpodndf jfias djal

giant garage

ELC www.elc.co.uk

stimulates senses | creativity | imagination | learning to talk

communication skills | fine motor skills | discover the world | physical development

social skills | problem solving | learning to read | instils confidence

hand to eye coordination | first words | gross motor skills | thinking skills

learning to write | learning to write

Icons have been introduced to help parents understand the specific skills that each toy can help their child develop—from imagination to fine motor skills.

PEN PORTRAIT

Thirty-plus moms and their children. These are relatively affluent parents who play an active role in their children's development. As a brand, ELC offers them the reassurance that their toys are designed to encourage learning through play—an idea implicit in the brand name. As well as having all-important child appeal, the toys offer the added benefit of helping with specific areas of growth and development.

A color-coding system is used for all packaging. The color palette for each product is dictated by the category of play within which it falls. Twelve categories have been defined in total. Here, the dark blue banding indicates "Action and Adventure" and also houses the ELC logo. High contrast orange makes the packaging impactful and acts as a foil for detailed images of the product.

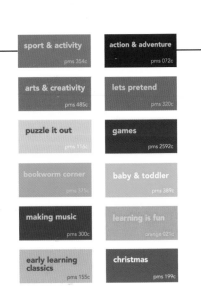

sport & activity
pms 354c

action & adventure
pms 072c

arts & creativity
pms 485c

lets pretend
pms 320c

puzzle it out
pms 116c

games
pms 2592c

bookworm corner
pms 375c

baby & toddler
pms 389c

making music
pms 300c

learning is fun
orange 021c

early learning classics
pms 155c

christmas
pms 199c

The packaging is made to work hard, with every surface used to convey information or offer inspiration to parents and children. Here, the side of the box highlights the play category and displays clear product features with close-up photography and yellow highlight boxes.

Playing tips have been incorporated into the packaging to give useful advice to parents on how to interact with the toy and their children. This maximizes both the child's enjoyment and the learning benefits of each toy—an essential ingredient in delivering ELC's brand promise: to help children be all they can be.

Ensuring child and parent appeal alike is second nature to Werner Design Werks. It created the brand Let's Eat for a Target private label kids' food concept. The packaging system was conceived to be fun for kids and communicate the nutritional attributes of the product to parents. Fun type and illustration work in harmony with food photography across a broad range of foods, from cereal to frozen chicken nuggets.

Occasionally, the children's category morphs into "kidult" territory. Sold primarily in boutique toy stores, Skwish is often purchased as an adult gift as well as a baby toy. Werner's proposed package redesign protects while allowing the customer to see and interact with the toy. The package needed to accommodate a lot of copy in several languages so the agency teamed up with writer Jeff Mueller to make it as fun as possible.

THIS PAGE AND
OPPOSITE BOTTOM LEFT:
Let's Eat range of food packaging
Design: Werner Design Werks

OPPOSITE TOP:
Skwish packaging
Design: Werner Design Werks

OPPOSITE MIDDLE
AND BOTTOM RIGHT:
Early Learning Centre Shopping Trolley
and Steam Iron packaging
Design: Parker Williams

KIDS_M/F
BRIGHT KIDS

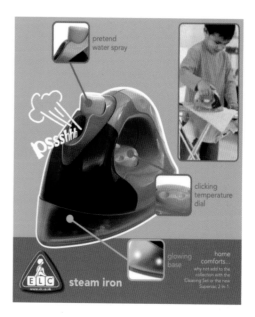

From the age of two onward, children become keen to be more self-sufficient. P&G's Kandoo is a range designed to help children by teaching them healthy habits for life. Research identified that kids are always saying, "I can do it myself,"—hence the empowering name, Kandoo, supported by the Pampers brand which offers parents quality reassurance.

The packaging is fun and specially developed with little hands in mind, so the pump is in the form of the frog's head, offering the practical solution of being easier for children's hands to push, and the bottle is designed to have a wide, stable base.

The vibrant green and purple color palette appeals to boys and girls alike and gives a fresh appearance. Through easy-to-understand graphics, the Kandoo frog shows that kids "can do" it on their own by demonstrating the proper elements of bathroom hygiene.

The three liquid products in the range—handsoap, bodywash, and shampoo—have different body shapes to help distinguish them, along with fun additions to the Kandoo frog's head, like soap bubbles for shampoo or a snorkel for the bodywash.

THIS PAGE AND OPPOSITE:
Pampers Kandoo range by P&G

KIDS_M/F
BRIGHT KIDS

Making learning fun is often a key objective of packaging in this sector, so type and illustration balance playfulness with information. Bright cheerful colors are appealing to kids while color coding aids a parent's purchasing decision.

Children's packaging design is becoming increasingly sophisticated. Boon products appeal to the design-conscious consumer looking for clean, modern design as an alternative to gaudy children's products and packaging. By accenting a simple background with quality photography and minimal text, the consumer gets all the information needed to make a purchase decision without being overwhelmed. Products are always shown in use and windows allow consumers to see and sometimes touch the product.

® Design has created packaging for Ladybird baby products and toiletries aimed at expectant and new parents. Spots express the logo over the entire packaging, with pastels on white used for baby products and the reverse scheme used to create brighter packs for toddlers.

Ladybird
Baby soap

Ladybird
Baby talc

Ladybird
Baby shampoo

Ladybird
Baby conditioner

Ladybird
Baby lotion

Ladybird
Kid's bubble bath

Ladybird
Kid's suncare

Ladybird
Kid's shampoo

Ladybird
Kid's conditioner

KIDS_M/F
BRIGHT KIDS

Frog Pod Deluxe
Bath Toy Scoop, Drain & Storage

Frog Pod & Bath Goods

10%

boon

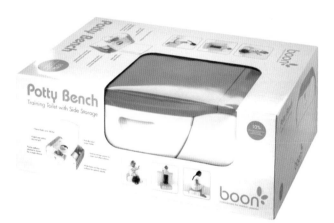

Potty Bench
Training Toilet with Side Storage

10%

boon

Sainsbury's
kids
Mini
organic
cherry
tomatoes

Keep refrigerated

Display until | Use by

nutritionally balanced range

Little monsters adorn Turner Duckworth's packaging for Superdrug's children's haircare range—a tongue-in-cheek reference to the little monster in every household. Fruity monsters hint at the fragrance of each product while others illustrate the product type: a two-headed monster for two-in-one shampoo and conditioner, a dripping three-eyed monster for after-swim three-in-one conditioning shampoo and body wash. Kids love the illustrations and the name connects with parents.

Brenda Lardner and Nick Bernard know a thing or two about keeping kids entertained. As joint directors of Easy Tiger Creative, they work with the Natural History Museum to create exciting exhibitions for families and children. Following a major rebrand, the museum approached them to bring some of the experience of the exhibits into the catering zones. Their witty solution assigns each sandwich variety an animal and uses terms like "herbivore" in place of the usual "vegetarian." The fun facts keep kids entertained before they dive back into the stimulating museum environment.

THIS PAGE:
Superdrug kids haircare range
Design: Turner Duckworth
Creative Directors: David Turner, Bruce Duckworth
Designer and typography: Sam Lachlan
Illustrator: Nathan Jurevicius

OPPOSITE:
Natural History Museum entertaining
and educational sandwich packs
Design: Easy Tiger Creative

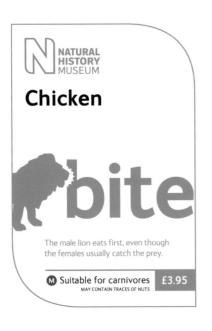

Chicken

bite

The male lion eats first, even though the females usually catch the prey.

Ⓜ Suitable for carnivores £3.95
MAY CONTAIN TRACES OF NUTS

Egg

chew

Pandas can eat huge amounts of bamboo, feeding for up to 12 hours each day.

Ⓥ Suitable for herbivores £2.95
MAY CONTAIN TRACES OF NUTS

KIDS_M/F
LITTLE CRITTERS

Tuna

snap

Crocodiles sometimes hunt together to herd fish towards them.

Ⓜ Suitable for carnivores £2.95
MAY CONTAIN TRACES OF NUTS

Cheese

nibble

House mice will nibble anything, from bread left out to soap in the bathroom.

Ⓥ Suitable for herbivores £3.50
MAY CONTAIN TRACES OF NUTS

Ham

gnaw

All adult members in a pack of wild dogs bring back food for the young.

Ⓜ Suitable for carnivores £2.75
MAY CONTAIN TRACES OF NUTS

NATURAL HISTORY MUSEUM

Monsters, dinosaurs, and living species all qualify for the Little Critters palette. Kids grow up learning about animals. The Critter typeface is a beautiful combination of education and entertainment—with each letter cleverly formed in the shape of an animal whose name the letter represents. Colors are bright and fun. Photography brings the species to life and offers an educational slant.

Easy Tiger Creative's packaging for the Natural History Museum's retail range brings the museum's "Power of Nature" positioning to life. Using the entire box as a canvas, each product features a full-bleed photographic image of the relevant species, creating powerful standout on the shelf and helping the NHM brand to "own" a section of the cluttered children's retail environment.

THIS PAGE:
Little critters palette

OPPOSITE TOP:
Early Learning Centre
Click Clack Dino pack
Design: Parker Williams

OPPOSITE BOTTOM:
Natural History Museum packaging
Design: Easy Tiger Creative

KIDS_M/F
LITTLE CRITTERS

CHAPTER 2 _KIDS _M

Why do boys and girls tend to prefer different toys? Scientists fall into two camps—those who believe it's down to genetic difference and those who believe that it's a product of socialization. It's the "nature versus nurture" debate. Moms and dads will often tell you a boy is more likely to use a stick as a sword than as a wooden spoon. Whatever your personal view, marketers seem pretty certain where to put their money. Action, adventure, space, and exploration are all selling points for products aimed at boys, and packaging designers use these themes as inspiration for eye-catching designs.

Fads and crazes in kids' toys come and go, but space has long been a perennial pull for little boys. Perhaps the key to its attraction lies in the versatility of the space theme: educational—astronomy and learning about the universe; adventurous—action-packed space travel; science fiction—*Star Trek* and all that has followed.

Davies Hall's rebranding and repackaging of the Doves Farm Organic brand cleverly leverages the space theme for its healthy yet tasty cereal, Chocolate Stars. With the astronaut sporting trendy aviator glasses, this is cool space travel with bags of little boy appeal.

One of Early Learning Centre's most successful ranges for boys is Planet Protectors, which brings an environmental slant to action and adventure. The Planet Protectors are a team of action heroes who protect planet Earth from the evil alien Doc Tox and his G.L.O.B.S. (Gross Lumps of Bad Smells), from making the world a smelly and polluted place. With strong sub-branding, the open packaging allows maximum interaction with the toy and encourages trial before purchase.

THIS PAGE:
Doves Farm Organic Gluten Free Chocolate Stars
Design: Davies Hall

OPPOSITE TOP:
Early Learning Centre's Planet Protectors space ship
Design: Parker Williams

OPPOSITE BOTTOM RIGHT:
Neutron Stunt Top for Science Museum
Design: Easy Tiger Creative

KIDS_M
ROCKET MAN

PLANET PROTECTORS

science museum

NEUTRON STUNT TOP *

* nothing stops this spinning top

Neutron Stunt Top

TOP SPINS AUTOMATICALLY WHEN PLACED ON PADDLE. SPINS ON ANY SURFACE, EVEN YOUR HAND

PROBABLY THE BEST TOP EVER INVENTED

5+

The space theme incorporates illustration and futuristic type to appeal to the little rocket man. A strong retro feel dominates—referencing 1950s atomic graphics and the optimism of the early space program. Werner Design Werks taps into this successfully with its cute Let's Eat packaging concept featuring a chicken-driven space ship.

Easy Tiger Creative's research gave them valuable insight into designing packaging for little boys. Working on a range for the Science Museum, they were convinced that simple packaging with one hero product picture would have real impact—white in a mass of visual shouting. Science Museum brand guidelines offered them a range of logo colors to work with. In-store research not only validated the appeal of this simple approach, but also indicated that magenta was the most powerful logo color—allaying fears that a pink logo would put boys off.

THIS PAGE:
Rocket man palette

OPPOSITE TOP:
Science Museum's Robot Bank
and Cosmic Rocket
Design: Easy Tiger Creative

OPPOSITE BOTTOM:
Chicken Nuggets from the Let's Eat range
Design: Werner Design Werks

KIDS_M
ROCKET MAN

CHAPTER 3 _KIDS _F

They may be destined to become lawyers, journalists, and doctors, but between the ages of three and eight, little girls only have one role model in mind: princess. Complete with tiaras, netting dresses, and, of course, princess slippers, this fantasy world has gathered momentum in recent years—not entirely without marketing impetus. Disney saw the opportunity to bring all its female heroines under one Disney Princess sub-brand, which now incorporates over 25,000 items. This lucrative democratization of princess status means every little girl can be one too.

Fantasy and fairy tales provide rich themes for products marketed to little girls. The phenomenon seems to cross cultures. The Chinese brand of children's toothpaste shown here features a fantasy figure ready to protect your teeth, complete with shield and toothbrush as weapon. The pink packaging gives clear gender cues and its innovative construction incorporates a separate stand to hold the tube upright. Its large toadstool cap gives it added fairy-tale appeal, while also being easier for little hands to twist off.

All that pink could be overpowering, but two examples of Disney packaging demonstrate that attention to detail pays off. Factor Design's packaging for the Disney television range uses each face of the box to evoke the magical world of the brand, using crisp product photography juxtaposed with beautiful illustration. Parker Williams makes the most of the princess theme, crafting the lids to the Magic Artist Dough range in the form of tiaras.

THIS PAGE:
Toothpaste in novel toadstool design

OPPOSITE TOP AND BOTTOM RIGHT:
Disney's Magic Art Dough
Design: Parker Williams

OPPOSITE MIDDLE RIGHT:
Cinderella TV packaging for Disney
Design: Factor Design

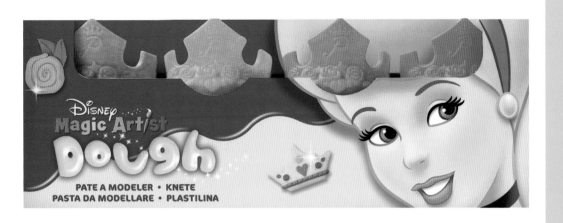

KIDS_F
PRINCESS

There's no escaping it, little girls are pretty in pink. Fairy tales offer rich territory, from the baroque excesses of an imaginary princess castle to mythological toadstools. Love hearts and homely gingham offer emotional warmth. Magical script type contrasts with clean and simple sans serif for clarity of information.

Another Chinese toothpaste uses a metallic pink finish to evoke a clean sparkly result, while natural flavor cues are offered through an elf-style character complete with leaf and fruit headgear.

Early Learning Centre's Cup Cake Rocking Cradle uses a rabbit cartoon and domestic gingham to denote love and care. Technology doesn't escape the pink treatment either. Every little princess can have a direct line to daddy with Firefly's love heart adorned cellphone.

THIS PAGE:
Princess palette

OPPOSITE LEFT:
Berry flavor toothpaste

OPPOSITE TOP:
Firefly phone casing in pink hearts

OPPOSITE BOTTOM RIGHT:
Cup Cake character and packaging design
for Early Learning Centre
Design: Parker Williams

KIDS_F
PRINCESS

CHAPTER 4 _TEENS/YOUNG ADULTS_M/F

With fewer children in the average household, 15–25-year-old consumers have access to more financial resources than ever before, either via their parents or their grandparents—a phenomenon the Japanese refer to as "six wallets." This group may be the focus of a great deal of marketing attention, but it's not easy to get right. The "No Logo" generation values the opinion of peers over corporate marketing. Successful packaging for these consumers reflects the experiences and attitudes of young urban cool-hunters, engaging their minds as well as their wallets.

CASE STUDY _LABEL M

"Specifically designed as a fashion label for hair, Label M helps Toni&Guy professionals interpret catwalk trends into wearable styles. The philosophy: simplicity, performance, and quality."

DAVID ROGERS

CREATIVE DIRECTOR, PureEquator

Briefed to create a global salon-exclusive haircare brand for Toni&Guy salons, PureEquator created the name, branding, and structural packaging for Label M. With strong competition from other professional brands, the range needed standout and desirability for a style-conscious consumer. "Toni&Guy has received both 'super brand' and 'cool brand' status," explains PureEquator's David Rogers. "The new range had to maintain the brand positioning without using the Toni&Guy brand name as [that] name is an exclusive retail haircare brand sold in Boots stores."

The name they created combines fashion (Label) with salon credibility (M for Mascolo, the surname of company founders Toni and Guy). PureEquator designed the world's first tooled bottle with a long-lasting rubber sleeve. Designed to be refillable, it also has environmental credentials. The urban-styled packaging with New York subway inspired graffiti graphics connotes carefree styling with dramatic results.

Revolutionary yet functional, the packaging has been a phenomenal success, both in terms of sales and awards, winning Gold in the Package Design category of the London International Awards.

OPPOSITE:
Label M for Toni&Guy
Design: PureEquator

label.m
label.m
CLEANSE + REPAIR SHAMPOO
LONDON
label.m
REPAIRING CONDITIONER
LONDON

Graphic designers Meat and Potatoes produce their own brand of clothing, prints, and books for young adult cool-hunters who shop in boutiques and online. With their T-shirts, they take the "limited edition" concept and keep it true to its artistic roots by creating a ready-to-frame hand silk-screened, numbered art print for each T-shirt.

The polystyrene meat tray packaging plays brilliantly on the company name. Creative Director and CEO Todd Gallopo sums up his design approach and concept thus: "Whenever I buy a T-shirt, it's most likely due to the art or graphic on it. Our concept of creating a limited-edition, silk-screen print of the same art that's on the shirt and custom packaging it with the T-shirt creates art that is both wearable and permanent." The "undesigned" style of the branding is information driven, allowing the T-shirt art to do the selling.

Clarity of information is also central to Werner Design Werks' branding for Outset, Target's private label brand of luggage. As an entry-price-point luggage range aimed at first time travelers and college students, icons included in the hangtag enable consumers to compare the different price points and features.

THIS PAGE:
Own brand T-shirt packaging
Design: Meat and Potatoes

OPPOSITE TOP:
Outset swing tickets
Design: Werner Design Werks

OPPOSITE BOTTOM:
Own brand T-shirt packaging
Design: Meat and Potatoes

**TEENS/
YOUNG ADULTS_M/F**
URBAN

The creation of designer/illustrator Jamie Hewlett and musician Damon Albarn, Gorillaz is the world's first virtual band. To mark the achievement of two platinum albums, Kidrobot released the 2-Tone edition set of figures. These are the final colorways of the three-dimensional vinyl creations with a run of just 1,000 sets. As every collector knows, retaining the distinctive packaging—white with spray paint logo indicating the color of the vinyl model—will be essential to maximize the set's future value.

THIS PAGE AND OPPOSITE:
The Gorillaz figures and packaging
Design: The Gorillaz and Kidrobot

TEENS/
YOUNG ADULTS_M/F
_URBAN

Designer Rune Mortensen used the urban landscape as the starting point for the CD cover design for *Seven* by Ken Vandermark and Paal Nilssen-Love. The cover uses a simple block of type to contain all the necessary information, much of which is usually considered back-of-pack copy. The inner sleeve features Rune's black-and-white photographs of moody, barely populated city scenes. In other projects, he lets typography carry the message.

The use of a gothic typeface on the cover of *Frozen by Blizzard Winds* by Kevin Drumm and Lasse Marhaug seems a strangely decorative choice. Used in conjunction with the prominent bar code, the juxtaposition creates a dynamic effect, combining the ornamental with the purely functional.

PureEquator's distinctive packaging for Label M also uses sharp contrast for impact. The sober colors of the containers contrast with vivid labeling inspired by the spray paint aesthetic of the graffitied urban environment.

THIS PAGE:
Inner packaging artwork for *Seven* by
Ken Vandermark and Paal Nilssen-Love
Design: Rune Mortenssen
Photography: Rune Mortenssen

OPPOSITE TOP:
Cover of *Seven* by
Ken Vandermark and Paal Nilssen-Love
Design: Rune Mortenssen
Photography: Rune Mortenssen

OPPOSITE MIDDLE:
Frozen by Blizzard Winds by Kevin Drumm
and Lasse Marhaug
Design: Rune Mortenssen

OPPOSITE BOTTOM:
Label M for Toni&Guy
Design: PureEquator

KEN VANDERMARK
PAAL NILSSEN-LOVE
SEVEN

1. First Hit, Second Fall (26:36)
2. Open Too Close (14:03)
3. Universal Funeral (3:19)

Ken Vandermark, tenor and baritone saxophone, Bb clarinet
Paal Nilssen-Love, drums and percussion

All compositions by Vandermark (ASCAP) and Nilssen-Love
(TONO/NCB). Recorded live at Blå on April the 1st,
mixed on December the 13th 2005 by Thomas Hukkelberg
at desibel.no. Live sound by Stig Gunnar Ringen. Produced
by Ken Vandermark and Paal Nilssen-Love. Co-produced by
Joakim Haugland. Photos and design by Rune Mortensen.
This recording is dedicated to Bjørnar Andresen.

www.smalltownsuperjazzz.com

SMALLTOWN SUPERJAZZZ

SMJZ

6 00116 84192   N℗B   STSJ119CD

℗© SMALLTOWNSUPERJAZZZ 2006

All rights of the
producer and of the owner
of the work reproduced
reserved. Unauthorised
copying, hiring, lending,
public performance
and broadcasting of this
record prohibited.

TEENS/
YOUNG ADULTS_M/F
URBAN

Frozen by
Blizzard Winds

Kevin Drumm
Lasse Marhaug

BARCODE 6 00116 83592 5

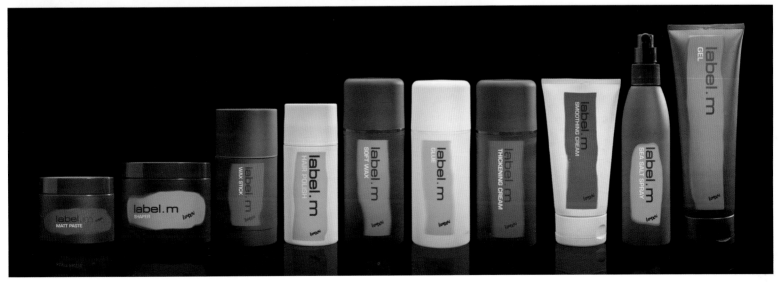

The urban environment offers graphic inspiration for packaging targeted at teens and young adults. The theme is gender-neutral and has an implicit cool factor. Representations of city scenes in mono- and duotones and materials such as rubber and raw kraft paper reflect the man-made aspect of the built environment. Typography mixes stencil, graffiti, and decorative faces such as gothic script. The colors are sober with vivid highlights.

Turner Duckworth's packaging for Motorola reflects the phones' design and functionality. Part phone, part MP3 player, the MOTOROKR's packaging features laser-cutting and the BurgoPak slider mechanism to create a dynamic three-dimensional speaker, communicating the music theme. Meanwhile the MOTORAZR's outer reflects the phone's distinctive tattoo styling.

B&W Studio created a new identity and clothing swing tags for clothing brand, Soul Cal. Swing tags emulate a graffiti artist's stencil: the bespoke logo is die-cut from raw cardboard, stickers applied, and a hole drilled to take an elastic band for fixing.

THIS PAGE:
Urban palette

OPPOSITE:
MOTOROKR and MOTORAZR packaging
Design: Turner Duckworth
Creative Directors: David Turner, Bruce Duckworth
Designers: Shawn Rosenberger, Ann Jordan,
Josh Michaels, Rebecca Williams, Brittany Hull, Radu Ranga
Structural Designers: BurgoPak, Turner Duckworth
Photo Illustration: Michael Brunsfeld (MOTOROKR)
Photography: Lloyd Hryciw (MOTOROKR)
Product imagery: Paul Obleas, Motorola

OPPOSITE MIDDLE AND BOTTOM RIGHT:
Soul Cal swing tickets and stencils
Design: B&W Studio

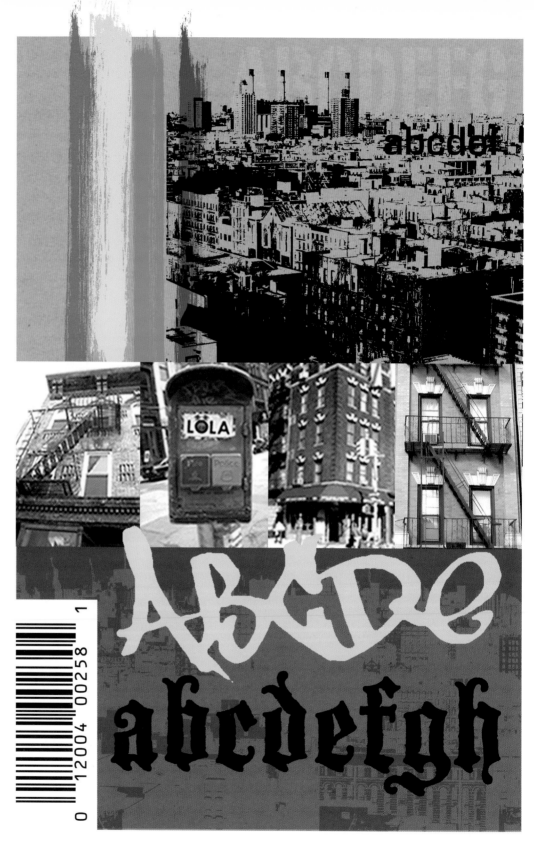

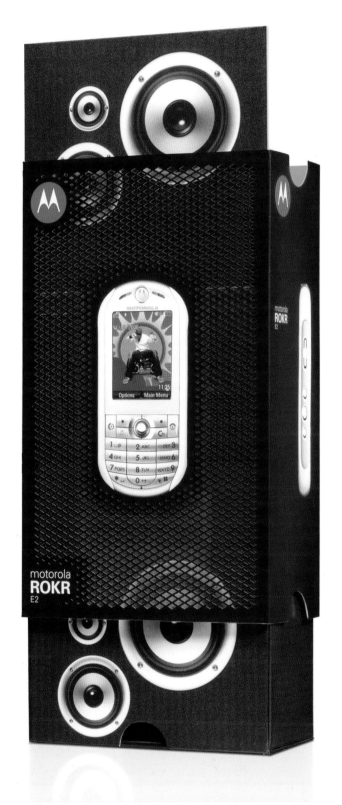

TEENS/
YOUNG ADULTS_M/F
_URBAN

CHAPTER 5 _TEENS/YOUNG ADULTS_M

According to a Euromonitor report, "The increasing time span between gaining financial independence and setting up a family is one significant factor in the burgeoning importance of young men as a consumer group." Delaying the need to grow up and take on responsibilities, this group focuses on having a good time and products and packaging use themes of hi-tech, performance, and energy to appeal to them. Marketers are capitalizing on the trend for young men experimenting with their appearance, spurred by sports and music celebrities and a plethora of magazines for men, from the Japanese *BiDan* (Beautiful Man) to *GQ*.

CASE STUDY _ANITA LIXEL

"The design is a direct response to the 'buy, rip, and archive' attitude created by iTunes. The target's music is all on their PC or iPod, while the source CDs are somewhere boxed in the attic, eBayed, given to friends, or thrown away. The target is more interested in buying individual tracks online and singles rather than complete CDs with artwork. Since selling music is evolving along FMCG (fast moving consumer goods) lines, I felt music should be packed as such—in throwaway packaging."

JIRI VANMEERBEECK

UTILIA

A chance encounter at a party between electro-pop singer Anita Lixel and designer Jiri Vanmeerbeeck proved a creative meeting of minds that led to innovative packaging for Anita's debut single, "In Your Game...Boy." "Utilia's process is very much hands-on and driven by gut feeling," explains Jiri. His innate understanding of Anita's fan base offered a starting point for the design, which needed to align with two key cultural themes: disposability and sharing.

As a limited edition of 500, the production budget was tight. The design consists of two transparent plastic CD trays, vacuum-sealed in a silver and transparent plastic bag. "Anita's look du jour was Edie Sedgwick," reflects Jiri, "so I took Andy Warhol's Silver Clouds installation as inspiration. The bag has a notched edge so you can easily tear it open, just like a candy bar."

The silk-screen printing on the CD was limited to two colors to save costs, and all information, including the bar code, is on the CD surface, "handwritten—with the charm of a demo tape."

OPPOSITE:
Anita Lixel
Design: Utilia

"As for sharing" continues Jiri, "to anticipate this in a positive way, each pack includes two identical CDs. The other one is to give away—so the packaging became a great vehicle for viral marketing. Some months later, a major label [in Belgium] followed this example."

Jiri knew that if Anita approved, her fans would too. "In this case the client was the stereotype of her audience—the i-Generation, computer-savvy, always online crowd. She absolutely loved the design on first sight."

Performance is a key theme in communicating with young men. Many brands make energy giving claims, and the theme has even crossed over into the alcohol sector. Created for the US market, Sparks is an alcoholic drink that also includes energy ingredients such as guarana, taurine, and ginseng. Working within tight government controls that would not allow the product to claim "alcohol and energy in a can," Turner Duckworth used visual communication to do the job. Its battery-inspired can gets the message across more potently than a product descriptor ever could.

Pack format is a key element in communicating a product's functionality. Taut is the world's first totally clean isotonic sports drink. Identica created a crisp pack design that communicates hi-tech through a silver finish, and natural refreshment through the fruit and water depiction.

Fuelosophy offers a softer, more natural take on the energy proposition. Launched by Pepsi with the proposition "rethink your energy source," it is their first natural and healthy energy drink. Templin Brink created the Fuelosophy brand and packaging, with its handcrafted logo and simple layout, to be distinctively different from Pepsi's regular mainstream grocery brands and appeal to the natural foods shopper.

THIS PAGE:
Sparks is an alcoholic beverage from McKenzie River Corporation, which also includes energy ingredients
Design: Turner Duckworth
Creative Directors: David Turner, Bruce Duckworth
Designers: Jeff Goeke, Allen Raulet, David Turner
Account Manager: Elise Thompson

OPPOSITE TOP:
Fuelosophy for Pepsi
Design: Templin Brink

OPPOSITE BOTTOM:
Taut Isotonic packaging
Design: Identica

TEENS/
YOUNG ADULTS_M
ENERGY

CASE STUDY _XBOX

"Gaming is like a drug. The ultimate quest for hero status. Many gamers are avid collectors—especially in Japan—so limited-edition merchandise appeals to them. Microsoft had experimented with a limited-edition console in green, but it hadn't really captured the gamer's imagination. For this special release, we needed to reflect current trends and position Xbox as the ultimate gaming machine."

PETER KRUSEMAN

MANAGING PARTNER, ENTERPRISE IG

In the gaming world, new technology is always round the corner. Having taken a substantial share of the market with the launch of Xbox, Microsoft was already developing the next generation of online gaming technology, subsequently launched as Xbox 360. With some gamers holding out for the launch, Microsoft looked to a tried and tested strategy used by the car industry to generate interest in current stock: the limited-edition release. They approached the Dublin office of Enterprise IG to come up with packaging that would keep the market buoyant until the new console was launched.

"There are three levels of gaming: Expert, Keen, and Occasional," explains Managing Partner, Peter Kruseman. "Expert gamers set the tone. They were our target market: urban males, generally 18–35-year-olds." The first thing that struck the team was the name. "Microsoft was intending to call the console Xbox Crystal," continues Peter. "We thought about the trend toward narrative-based action and adventure gaming and felt that 'ice' had strong connotations. It's evocative of stories where the key to the mystery is frozen in ice. So we proposed an icy feel to the Crystal pack to give it more resonance with gamers."
→

OPPOSITE:
Xbox special-edition packaging
Design: Enterprise IG
Creative Director: Simon Richards
Designer: Sarah Maguire

→ In addition, the team felt that two key brand attributes were being underplayed in current packaging. "The green jewel had always been a key element of the brand language. It suggests a promise of what lies beyond," says Peter. "Also, the X is a compelling brand attribute that suggests intrigue, like the *X Files*. We brought the console back to be the hero of the packaging, supported by these key brand elements."

The results impressed Microsoft so much that what had started as a remit for Europe, the Middle East, and Africa was implemented globally and the Enterprise IG team was invited to redesign the other packs to bring them in line: the original black console (referred to as the Vanilla pack) and bundled console and game promotions, which were an important part of the sales strategy. The new pack architecture created a more coherent and eye-catching visual identity across all three products and a stronger brand language that connected with the target audience.

THIS PAGE AND OPPOSITE:
Xbox packaging
Design: Enterprise IG
Creative Director: Simon Richards
Designer: Sarah Maguire

Energy is communicated through a combination of hi-tech, performance, and speed. Silver, black, and white are all used as clean bases for pure color, which offers a technical edge. Type is a combination of futuristic and blocky faces with italicized type conveying a sense of speed and movement. Symbols such as positive and negative signs refer to electrical current and three-dimensional button effects suggest interactivity.

Portable energy is the idea behind USB Cell: AA batteries that recharge via a concealed USB connection. The logo incorporates a graphic representation of the existing "B" shape found within the connector of a standard USB plug. A shot of lime green gives both product and packaging instantly recognizable visual punch.

The palette can also take on an edgy feel. Turner Duckworth created an unholy mix of snowfakes, warning signs, and Ninja throwing stars on metallic foil pouches to convey the "dangerously cold" proposition of Zero Degrees, a frozen soft drink with attitude.

THIS PAGE:
Energy palette

OPPOSITE:
USB Cell brand identity and packaging
Design: Turner Duckworth
Creative Directors: David Turner, Bruce Duckworth
Designer: Jamie McCathie
Typographer: Jeremy Tankard
Retoucher: Reuben James

OPPOSITE BOTTOM:
Zero Degrees for McKenzie River Corporation
Design: Turner Duckworth
Creative Directors: David Turner, Bruce Duckworth

AA
Ni-MH

AA rechargeable batteries
Plug into any usb port to recharge

USBCeLL
moixa

No charger needed

USBCeLL
moixa

Simple, re-usable power

USBCeLL
moixa

USBCeLL moixa

USBCeLL moixa

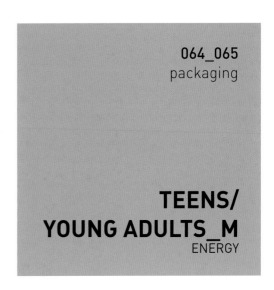

064_065
packaging

TEENS/
YOUNG ADULTS_M
ENERGY

MSRP
99¢

zero°™
FROZEN COLA DRINK

8FL.OZ. (237ml)

MSRP
99¢

zero°™
FROZEN CHERRY DRINK

8FL.OZ. (237ml)

MSRP
99¢

zero°™
FROZEN GRAPE DRINK

8FL.OZ. (237ml)

MSRP
99¢

zero°™
FROZEN BERRY DRINK

8FL.OZ. (237ml)

While the men's fitness and grooming category goes from strength to strength, 15–25-year-old men are still cautious consumers and look for reassuringly masculine cues in packaging to ensure that a product is intended for them. The Men Only grooming and styling range, designed by PureEquator for Toni&Guy, uses simple matte black structures and electric blue neon signage to communicate the tongue-in-cheek message. The range stands out in a sea of grooming products targeted at this group with a professional styling appearance and a confident positioning.

Niche brands are looking for ways to broaden their appeal in the health and fitness category. Identica helped reposition MMUSA as an innovative and progressive expert in the growing sports supplement market, while not alienating its core market of the "hardcore" body builder. An extended range of products focuses on helping athletes achieve their personal best. The pharmaceutical label style (square to reference the proprietary serum bottle) and silver bottle color were chosen to evoke scientific efficacy. A color palette for each of the product categories helps consumers identify products easily: red for power, green for fitness, blue for recovery, yellow for health and vitality, and purple for sex-enhancing products.

THIS PAGE:
Toni&Guy Men Only product packaging
Design: PureEquator

OPPOSITE:
MMUSA packaging range
Design: Identica

**TEENS/
YOUNG ADULTS_M**
MEN ONLY

MARVELLUS

NATURAL ENERGY
ENHANCING SERUM

FUEL
E · F

5.1 FL OZ | CHERRY

MUSCLE MARKETING USA

ANTI-OXIDANT
SERUM

> Accelerates Recovery
> Combats Free Radicals
> Prevents Cell Damage
> No Side Effects

ANTI-OX
RECOVERY

MUSCLE REPAIR FORMULA

♂ MALE | 5.1 FL OZ | CHERRY

MUSCLE MARKETING USA

RUNNERS SERUM

> Builds Strength
and Stamina
> Defeats Fatigue
> Protects Joints

ENDURUS
FITNESS

MALE ENDURANCE FORMULA

♂ MALE | 5.1 FL OZ | GRAPE

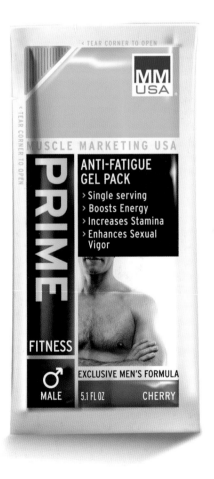

< TEAR CORNER TO OPEN

< TEAR CORNER TO OPEN

MM
USA

MUSCLE MARKETING USA

PRIME

ANTI-FATIGUE
GEL PACK

> Single serving
> Boosts Energy
> Increases Stamina
> Enhances Sexual
Vigor

FITNESS

♂ MALE | EXCLUSIVE MEN'S FORMULA

5.1 FL OZ | CHERRY

The straightforward and masculine communication style features black as a dominant color with all shades of silver and gray offering a technical edge. Cool blue, vibrant green, and sharp orange provide highlights. Contemporary type is strong and structured sans serif, with the occasional witty touch like neon signage.

A recent Euromonitor International report stated: "Chinese consumers are showing an increased preoccupation with personal appearance and image, as well as greater interest in Western brands or norms, suggesting that China is ripe for further growth in cosmetics and toiletries." Western brands are advised to be sensitive to cultural nuances. Mentholatum displays an understanding of the consumer while also offering products and packaging with standout design. Its range of lip balms includes a specific offering for men, packaged with strong masculine performance cues. A range of body washes with natural extracts has employed a two-tier communication strategy. The sober colors and type indicate a masculine offering, but to give further reassurance to the male consumer, removable stickers depict Chinese men using the products.

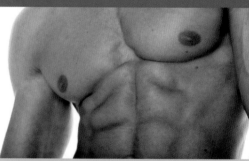

THIS PAGE:
Men only palette

OPPOSITE:
A collection of Mentholatum products for men, designed for the Chinese market

**TEENS/
YOUNG ADULTS_M**
MEN ONLY

CHAPTER 6 _TEENS/YOUNG ADULTS_F

The media tends to represent two extremes of this group: the pampering babe or the party girl. Somewhere in between is a savvy, independent minded, high achieving young woman with unprecedented spending power. The subtleties and complexities involved in designing successful packaging for teen girls and young women should not be underestimated. Fashion, beauty, technology, health, and indulgence products have to demonstrate that they can empower, excite, and involve these consumers. Glamor, playfulness, irreverence, and seduction all play a part in making the vital connection.

CASE STUDY_SECRET WEAPON

"Research had identified the need for a range of toiletries and cosmetics that would actively appeal to 13–18-year-old girls, but not their mothers. In particular, it was important to use physical packaging that was totally different to the packs bought by their mothers. Hence the use of blood bags to contain the bath soaks—cool to this age group, but slightly off-putting to their mothers."

BRUCE DUCKWORTH

CREATIVE DIRECTOR, TURNER DUCKWORTH

Developed for drugstore chain Superdrug, the Secret Weapon range by Turner Duckworth used consumer insight to create cut-through packaging design that really hits the spot with its target audience. The range is aimed at teen girls who have either a weekly or monthly allowance. They are experimenting with their appearance and change their fragrance and the color of their eye shadow regularly, which meant the low price per item was key for the success of the range.

Research was used judiciously to inform each stage of the design process. Initially, qualitative groups were conducted with the target audience. These allowed the Turner Duckworth team to identify key likes and dislikes of products currently on the market. Groups were used to validate the brand proposition and positioning, and then again to gain insight on the brand identity and packaging concepts.

The designers selected all the packaging materials for their tactile qualities as well as newness to the toiletries sector. For example, the solid perfumes were packaged in small, brushed-metal tins—the perfect size for slipping into a jeans pocket—with the lid embossed with the Secret Weapon identity. Wherever possible, the fragrance-indicating product colors were allowed to show through transparent packaging to create as much drama on shelf as possible.

OPPOSITE:
Secret Weapon for Superdrug
Design: Turner Duckworth

Following the launch, customer research groups indicated a high level of brand awareness among teenage girls during the first year of sales, with encouraging feedback such as: "It gives you the edge over your friends to get boys...It's different...It gets you noticed."

Respondents clearly felt that the brand was "for them" and positioned as something different. The figures tell the whole success story. The range was Superdrug's most successful launch ever. Initial target sales had been set at £1.5 million for year one. Actual sales were £3 million.

Secret Weapon
temptation
shower power

Secret Weapon
energy
shower power

Secret Weapon
purity
shower power

Secret Weapon
desire
shower power

TaB Energy is a fruity, pink, low calorie energy drink created by the Coca-Cola Company to offer stylish young women the energy to multitask. Turner Duckworth took up the challenge of reinventing this classic brand that had enjoyed popularity in the 1970s and 1980s. With the target audience in mind, it looked to the world of fashion to create an identity system based around a distinctive pattern more than a logo. The pattern is based on an optical illusion in which dots seem to appear and disappear as you look at it, thus giving the structured design a distinctive expression of energy, supporting the brand promise: Fuel to be Fabulous™.

Optically vibrant patterns feature in a number of packaging concepts intended for this consumer group. Multicolored squares, referencing iconic 1960s Pop Art feature in Pop Ink's Repeat Engagement note cards. Mentholatum's Lip Ice is a range of flavored lip balms popular with young Chinese women. The packaging features fun icons and retro color blocks, which help to depict flavor. Also for the Chinese market, Youngrace hair gels use bright funky colors and graphic shapes to reflect the differing strengths of the gels: a splash for water based, spiky for extra hold.

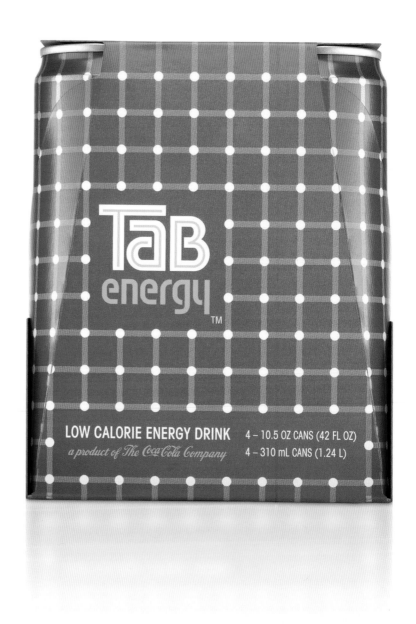

LOW CALORIE ENERGY DRINK   4 – 10.5 OZ CANS (42 FL OZ)
a product of The Coca-Cola Company   4 – 310 mL CANS (1.24 L)

仕甲冰
[西柚]
grapefruit

養秀雷敦

lip Ice

LipIce Lip Pot
Refreshing Grapefruit
Contains Jojoba Oil,
Olive Oil & Vitamin E.
Relief for dry,
chapped lips. Keep
lips soft and
Apply as nee

The Mentholat

elMentholatum
To LipIce Lip Pot
Night time
lip moisturizer.
Apply before bed.

lip Ice
NTHOLATUM

仕甲冰
[柠檬]
lemon

養秀雷敦

lip Ice

Lip Balm.Sunscreen
Protectant. Cooling,
refreshing, relief for
dry, chapped lips.
Apply as needed.
SPF15
The Mentholatum Co.

Mentholatum
LipIce Lipbalm
SPF15

TEENS/
YOUNG ADULTS_F
POP

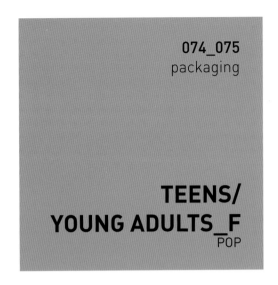

NOTE CARDS

BLANK

Pop ink

16 BLANK NOTE CARDS & ENVELOPES

POP INK · MODERN ARTIFACTS FOR TODAY'S POP CULTURE

Pop ink DESIGN

YOUNGRACE 温雅

HAIR GEL
温雅定型啫喱膏
特强定型
Extra-Hold
Formula

净含量200ml

YOUNGRACE 温雅

HAIR GEL
温雅定型啫喱膏
保湿亮泽
Water Based
Formula

净含量200ml

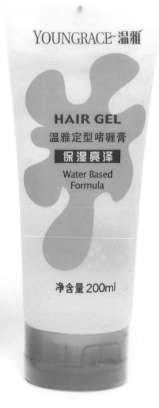

FOCUS _TANTRIC TONIC

The Partners created an identity for this smoothie bar that marks it out as different from most healthy offerings. The relatively simple typographic logo treatment is combined with a distinctive psychedelic swirling pattern in juicy retro pinks, violets, and yellows. The brand reflects the owners' passion for all things healthy, while steering clear of the usual juice bar health cues. Instead, carrier bags display the profiles of two faces (about to kiss?) and point-of-sale materials display strawberries oozing juice coupled with suggestive language about their aphrodisiac qualities.

PEN PORTRAIT

Tantric Tonic is a funky juice and smoothie bar catering for young office workers leading a hectic credit/debit lifestyle. Ideally situated for a delicious quick fix of goodness before the office day starts, Tantric Tonic's infectious passion for smoothies and juices prompted one online reviewer to write "I think that everyone should be forced to go to Tantric Tonic for a detox every Monday morning. They are fab!"

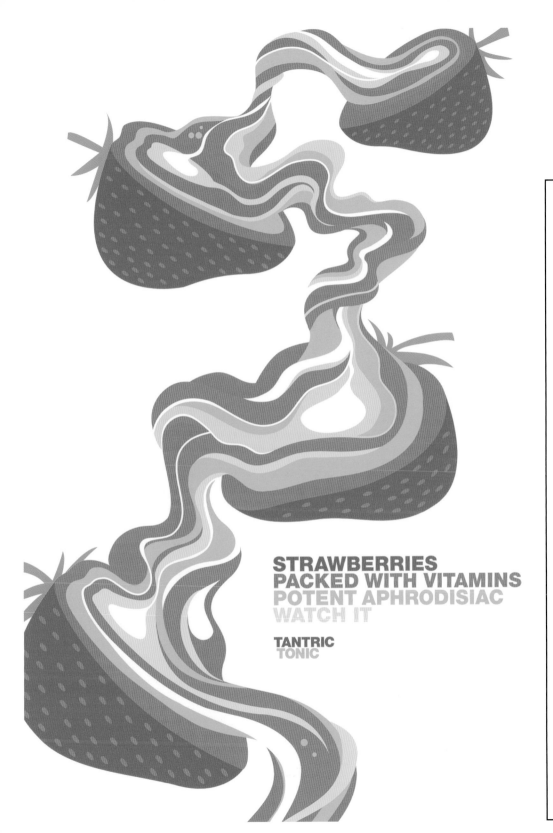

**STRAWBERRIES
PACKED WITH VITAMINS
POTENT APHRODISIAC
WATCH IT**

**TANTRIC
TONIC**

The psychedelic pattern helps to bring together different communication elements to tell a coherent brand story, whether adorning a point-of-sale poster or graphically representing smoothie goodness being sucked up through a straw on the takeaway cup.

Field's Sweet Spot is a nostalgic candy destination in department store Marshall Field's. With a nod to the 1960s' Mod style, creative agency Wink created branding, retail environments, and packaging for the concession with a vernacular that's equal parts fashion and fantasy. Wink describes it as "a concoction that combines both the department's retro-candy product selection and fashion driven branding efforts of Marshall Field's itself."

The identity and packaging creates a playful pop aesthetic that appeals to both children and Marshall Field's core female consumer alike. Fun circles adorn gift packaging that looks like elegant circular hat boxes, continuing the fashion theme. Selection boxes of chocolates carry the circular pattern in sumptuous purples and violets. Chocolate bars carry witty, occasion-specific messages, encouraging informal sharing.

THIS PAGE AND OPPOSITE:
A selection of packaging for
Field's Sweet Spot for Marshall Field's
Design: Wink

**TEENS/
YOUNG ADULTS_F**
¯POP

PURE MILK CHOCOLATE
THINKING OF YOU
Aw, isn't that sweet

PURE MILK CHOCOLATE
HAPPY BIRTHDAY
confection celebrating your conception

PURE MILK CHOCOLATE
THANK-U
reward for your thoughtfulness!

PURE MILK CHOCOLATE
CONGRATS
Go ahead, spoil your dinner, you've earned it

Circles abound in the pop theme: friendly, feminine circles; retro, funky circles; dynamic circles. Absolute Zero° created the name, identity, and packaging for fashion accessories brand, Mine Successories. Designed to appeal to bright young female consumers throughout Europe, the identity uses circles to create a strong brand feel and offers the packaging a useful window to display product.

Hot pinks, violets, and lilacs are teamed with fruity, zesty brights. Davies Hall created the name and brand style for Profusion, a probiotic fruit shot in four flavors. Merging health, beauty, and fashion styles, the packaging features dynamic circles in a juicy palette to appeal to a younger, health aware audience.

Type is rounded and retro or incorporates jaunty movement. Illustration references nostalgic glamor, epitomized by ® Design's packaging for the Glam makeup range. Targeted at 16–24-year-old women, this Woolworths brand uses imagery influenced by 1940s and 1950s advertising to create engaging, lighthearted packaging.

THIS PAGE:
Pop palette

OPPOSITE LEFT:
Mine Successories branding and packaging design
Design: Absolute Zero°

OPPOSITE TOP:
Profusion probiotic fruit shots
Design: Davies Hall

OPPOSITE MIDDLE:
Glam packaging for Woolworths
Design: ® Design

OPPOSITE BOTTOM LEFT/RIGHT:
Mentholatum's Lip Ice strawberry lip balm packaging
Coca-Cola TaB Energy can
Design: Turner Duckworth
Creative Directors: David Turner, Bruce Duckworth
Designers: Sarah Moffat, Chris Garvey

TEENS/
YOUNG ADULTS_F
POP

mine
successories

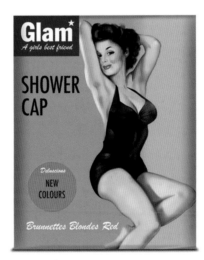

Glam*
A girls best friend

SHOWER
CAP

Delicious
NEW
COLOURS

Brunnettes Blondes Red

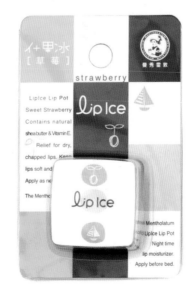

strawberry

lip Ice

LipIce Lip Pot
Sweet Strawberry
Contains natural
shea butter & Vitamin E.
Relief for dry,
chapped lips. Keep
lips soft and
Apply as ne

The Menth

lip Ice

Mentholatum
LipIce Lip Pot
Night time
lip moisturizer.
Apply before bed.

TaB
energy.™

a product of The Coca-Cola Company
LOW CALORIE ENERGY DRINK
10.5 FL OZ (311 mL)

CHAPTER 7 _ADULTS _M/F

Style obsessed DINKYs—"Dual Income, No Kids Yet" as marketers once labeled them—and their "Pink Dollar" gay counterparts are united from Mumbai to New York by above average disposable income with which to furnish their contemporary urban dwellings. More likely to spend time entertaining at home, they purchase premium food in upscale shopping environments, or splash out on luxury gifts. With their hard partying days behind them (well, almost), they're more discerning about the origins and health benefits of the food they eat—especially once little ones come along. When developing packaging for this sector, designers must create appeal for men and women alike. This calls for a neutral, but by no means impersonal, design palette.

Metropolitan style is understated, clean, and modern—exemplified by Aloof Design's packaging for Robin and Lucienne Day's reissued homewares for twentytwentyone. The packaging couples recycled kraft paper with silk-screened graphics to complement the natural product materials of wood and linen. A strong geometry prevails, with circular "portholes" that reveal the product inside.

Our obsession with the sleek lines of mid-century modern furnishings has had a direct influence on the way we furnish our homes. The style has filtered through to graphics and packaging too, with purveyor of contemporary furnishings West Elm, embracing 1950s-styled silhouettes on its carriers designed by Templin Brink. The redesign saw a huge increase in brand loyalty and sales.

For this target market, what you leave out is as important as what you put in. Three examples that typify this approach are: Turner Duckworth's simple color swatch on stainless steel solution for Homebase's sophisticated paint range, The White Room; ® Design's bold typographic approach to Kelly Hoppen's interiors range for Bhs; and the monotone print and emboss of Aloof's packaging design for Hush.

THIS PAGE:
Lucienne and Robin Day
packaging for twentytwentyone
Design: Aloof Design

OPPOSITE TOP LEFT/RIGHT:
The White Room for Homebase
Design: Turner Duckworth

OPPOSITE MIDDLE:
KH Home identity and bags for Bhs
Design: ® Design

OPPOSITE BOTTOM LEFT/RIGHT:
Hush scented candle packaging
Design: Aloof Design
Bags for West Elm
Design: Templin Brink

SOFT VANILLA WHITE

HOMEBASE

THE WHITE ROOM®

A durable, wipeable emulsion for interior walls and ceilings

MATT EMULSION

HOMEBASE

SOFT THISTLE

THE WHITE ROOM®

A quick drying, low odour, mid sheen paint for interior wood and metal

EGGSHELL

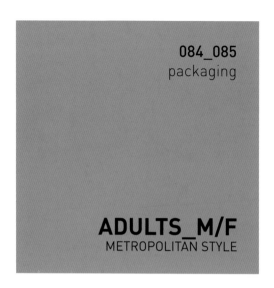

ADULTS_M/F
METROPOLITAN STYLE

KH_OME
KELLY HOPPEN

scented candle

www.hush-uk.com

hush

west elm

west elm

FOCUS _SIMPLE HUMAN

Understanding the target consumer, Smart Design determined the simplehuman brand strategy: to shift the focus from product to lifestyle—how you live your life and how the quality of your life is improved with these products. A dynamic packaging system conveys a consistent message by emphasizing benefits, rather than features, using friendly text, clean design, and easy-to-read graphics.

PEN PORTRAIT

simplehuman is a brand aimed at busy adults who want to take the stress out of daily tasks. These homewares are designed to improve efficiency at home—from a spoon rest on a utensil holder to a dishrack that drains in different directions. The ingenious design of these products is matched by sophisticated contemporary styling that appeals to modern urban homeowners.

When Smart Design began working with simplehuman, the company was named Can Works, Inc. They needed a new brand identity and packaging system that would enable them to expand beyond trash cans into new product areas. Repositioning the brand to encompass a lifestyle approach, Smart Design developed the simplehuman name, brand, and personality. This has helped simplehuman expand into five new product categories and grown the company by 30 percent.

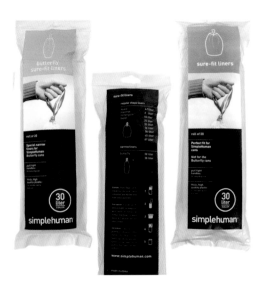

The simplehuman philosophy is based on providing "tools for efficient living®." The graphic style supports this with a combination of lifestyle and product photography, linear icons to denote the benefits of each product, and a color-coding system to indicate product style. This is repeated across retail and communications ensuring simplehuman products stand out on store shelves and helping customers to make an informed choice in a busy store environment.

Urban professionals who lead fast-paced social and work lives have little time to cook. Increasingly knowledgeable about food, they are heavy users of restaurants and ready-made gourmet meals. Wally's Food Company offers products that recreate the intimacy of a gourmet meal prepared at home by a professional chef.

Philippe Becker Design created a sophisticated brand identity with an appetizing color palette to delineate each of the four menu categories: soups and starters, entrees, sides, and desserts. With the food hand-packaged in clear, stand-up pouches, a double-sided label was developed to attach at the top. One master label works for every item in a category through the use of add-on laser-printed stickers.

Seeking an antidote to hectic lifestyles, adult consumers increasingly look to the East for traditional products like Japanese matcha tea, introduced by the Muzi brand. Aimed at culturally engaged urban sophisticates, Muzi is positioned as the next staple in contemporary world culture. The packaging supports this using intense, dynamic color and texture to denote individual tea families and a clean, Japanese aesthetic. Since these consumers expect authentic offerings, the Muzi brand puts the emphasis on educating customers on the tea brewing process, the quality of tea varieties, and the history of tea.

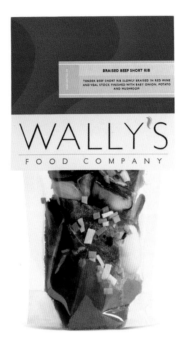

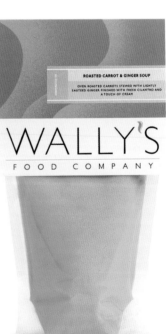

THIS PAGE:
Wally's Food Company packs
Design: Philippe Becker Design

OPPOSITE:
Muzi green tea packaging
Design: Identica

ADULTS_M/F
METROPOLITAN STYLE

Fusion products are part of a trend that appeals to this consumer. The Umi range, designed by Pearlfisher for high-end supermarket Waitrose, uses elements of premium food packaging to create a "gourmet toiletries" theme, helping to transfer the store's food expertise to its non-food ranges. Japanese for "beauty," the name Umi evokes a sophisticated marriage of beauty and food for the skin. The minimal design uses simple shapes and finishes. The premium black labels feature tempting names like "body soufflé" in punchy hot pink typography with exotic ingredients lists that convey a sense of luxury, provenance, and perfection.

Department stores naturally have to have a broad appeal. With heightened levels of sophistication and a competitive market in which the range of brands on offer in-store is crucial, The Partners have redesigned the House of Fraser identity. The new identity features a contemporary and pared down logotype, which appears in a series of contemporary colorways. The result is an upmarket and stylish feel that supports rather than competes with the other brand identities under the same roof.

THIS PAGE:
House of Fraser identity and bags
Design: The Partners

OPPOSITE:
Umi for Waitrose
Design: Pearlfisher

ADULTS_M/F
METROPOLITAN STYLE

peach kernel & vanilla
exfoliating body wash
for glowing skin

white rice & wheatgerm
bath soak to relax
mind & body

lotus flower &
oat ceramide
face cream
for radiant skin

vanilla & brown sugar
bath grains to
soothe mind & body

vanilla & cardamom
body soufflé
for velvety skin

umi

umi

umi

umi

Packaging for the urban sophisticate often features a subtle color palette with shots of vibrancy, exemplified by jkr's use of warm tones coupled with fruity shades for the Molton Brown range. Pared down design is key to the look. Bold sans-serif type works in a combination of weights. Simple line drawings and bold icons are used in favor of photography.

Metallic finishes and transparency reflect the materials of the urban environment, or create a restrained sense of luxury, as exemplified by Yang Rutherford's C-Side Spa range for Cowley Manor Hotel.

THIS PAGE:
Metropolitan style palette

OPPOSITE TOP LEFT/RIGHT:
Cowley Manor, C-Side Spa product packaging
Design: Yang Rutherford
Muzi tea packaging
Design: Identica

OPPOSITE BOTTOM:
Molton Brown gift packs
Design: jkr
Identity design and store carriers
Design: The Partners

ADULTS_M/F
METROPOLITAN STYLE

CASE STUDY _CHARLES CHOCOLATES

"We positioned this new brand to stand out in a saturated market through a sophisticated style combined with a good serving of whimsy."

JOEL TEMPLIN

CREATIVE DIRECTOR, TEMPLIN BRINK

Chuck Siegel has been a part of the San Francisco chocolate scene since 1987 when he started his first premium chocolate company, Attivo Confections, at the age of 25. For over seven years, Chuck dedicated himself to "recreating many of our childhood favorites with a decidedly gourmet sensibility."

By setting up his own brand, Charles Chocolates, Chuck was entering the traditionally conservative high-end confections market. He commissioned Templin Brink to create the brand identity and packaging for this very personal venture. Dedicated to using only the finest ingredients, including some of the world's best chocolates, organic herbs, fruits, and nuts, the brand needed to connote the product's premium status while also expressing the individuality of its founder.

The handdrawn logo and series of patterns reflects the handmade quality of the chocolates. The richness of chocolate brown is offset by contemporary silver and a palette of approachable colors. The result is a modern, understated design architecture that connects with its discerning consumer, helping the brand to achieve tremendous visibility from day one.

THIS PAGE:
Charles Chocolates identity and packaging
Design: Templin Brink

A couple of decades ago, design rules were broken as black became the norm for premium food packaging. Previously, black had always been considered unappetizing; now it connotes premium and gourmet. When ® Design redesigned the packaging for Selfridges, they took this insight and made it their own. "Food Noir," as they've playfully dubbed it, is based on the strong principle of one color, one typeface, one point size. The range is striking, coherent, and yes, appetizing.

Briefed to shift the Green & Black's brand from niche to an exclusive yet approachable level, Pearlfisher used their future insight research, TasteMode, to unlock the meaning of organic. They realized it would soon no longer just mean worthy; it would mean premium and importantly, tasty. The dark brown color clearly communicates intense flavor first, while the gold typography of the logo acts as a cue to the brand's premium status.

Metallics help to indicate premium too. Hence, Davies Hall used a single die-cut hole in the packaging for Bouchon D'Argent (meaning "silver cork") to reveal the silver of the product sharply against the black and rich red of the packaging and graphics. Targeted at the design savvy Japanese gift market, Mayday's design for Nairobi Coffee makes use of a distinctive tin and a sophisticated color palette supporting a single-minded mnemonic "N." High-quality production values, notably the contrast of matte versus metallic finish, contribute to the premium presentation. It's a memorable solution, which deliberately avoids the usual coffee language clichés.

THIS PAGE:
Charles Chocolates identity and packaging
Design: Templin Brink

OPPOSITE TOP:
Nairobi coffee packaging
Design: Mayday

OPPOSITE MIDDLE LEFT/RIGHT:
Bouchon D'Argent corkscrew
and bottle stopper packaging
Design: Davies Hall
Green & Black's chocolate wrappers
Design: Pearlfisher

OPPOSITE BOTTOM LEFT/RIGHT:
Selfridges food packaging
Design: ® Design

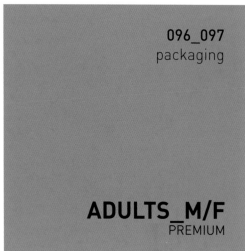

BOUCHON

SILVER PLATED BOTTLE STOPPER

SILVER PLATED BOTTLE STOPPER

DARK 70%

GREEN &BLACK'S

ORGANIC

DARK CHOCOLATE WITH 70% COCOA SOLIDS
100g ℮

WHITE

GREEN &BLACK'S

ORGANIC

CREAMY VANILLA WHITE CHOCOLATE
100g ℮

MILK

GREEN &BLACK'S

ORGANIC

A DARKER SHADE OF MILK CHOCOLATE
100g ℮

BOUCHON D'ARGENT

BOUCHON D'ARGENT

SELFRIDGES

CHARDONNAY VIN DE PAYS D'OC 2001

SELFRIDGES

MERLOT VIN DE PAYS D'OC 2001

SELFRIDGES

PURE GROUND KENYAN GETHUMB-WINI COFFEE

sharp, bright, invigorating strength 5

SELFRIDGES

MEDIUM CUT ORANGE MARMALADE

SELFRIDGES

RUNNY HONEY WITH HONEYCOMB

FOCUS _WORLD OF FOOD

The packaging concept was developed to work across 300 SKUs (stock keeping units) and a variety of food products ranging from wine and champagne to chocolates, cookies, smoked salmon, pasta, and canned foods. For ease of implementation, the mark had to work clearly in black and white.

PEN PORTRAIT

Department store House of Fraser has successfully embraced the trend toward "masstique"—products that are mass designed, but with an aesthetic edge that feels luxurious. With its World of Food concept, it targets adults interested in the premium delicatessen market, offering them products at a more accessible price point, with a more relaxed attitude.

The brand reflects a modern approach to food and cooking using design cues that are immediately recognizable as premium, softened with a warmth and approachability that speaks to the House of Fraser consumer. The World of Food sub-brand uses a modern, simple, uniquely drawn font style with the word "food" highlighted to communicate the offering with clarity and confidence.

Representing food produced around the world, the photographic concept features products highlighted in color against a black-and-white background. The same photographic concept is used in the food-hall retail environment. This approach cleverly combines the refinement of black and white (the premium category norm) with the appetite appeal of color photography. Where packaging features a clear window, the visible product takes the place of color photography.

Busy urban lifestyles coupled with increased demand for quality foods have given rise to the gastrodrome phenomenon. Foodstore, restaurant, and bar in one, Villandry exemplifies this type of destination, where customers are as likely to meet for after-work drinks as a three-course dinner. Or, buy the components to make a great meal in minutes at home. Courtesy of Dalziel + Pow, the packaging of its own label foods is as impeccably on target as the retail concept.

Two examples from Identica bring personality to the conventions of premium packaging. Based in London's fashionable Notting Hill, a local business approached the international branding experts to create an identity for its venture: an upscale coffee house and grocery store just round the corner. The identity for Kitchen and Pantry features a witty take on a mug tree, executed across packaging in appetizing earthy tones and silver. Another inventive solution was created for the Urban Garden Honey Co. Bees make honey from flowers; but the flowers on this jar of honey are all beautifully photographed stone-carved examples from the streets that surround the company's beehives.

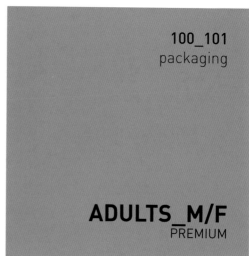

ADULTS_M/F
PREMIUM

For communicating premium food, black, silver, and rich colors prevail. The varying proportions of this palette help to bring differentiation. In the case of the Maitre D' range, named and branded by Davies Hall, silver evokes a clean surface deli feel, while Parker Williams' solution for Sainsbury's upscale range, Taste the Difference, steers clear of black in favor of a consistently employed branding device in rich purple to pull together premium products from diverse categories.

Black-and-white photography offers a sophisticated slant on provenance, while clean sans-serif typography can present a contemporary approach, as witnessed in The Partners' bold work for Klein Caporn's pasta sauces.

THIS PAGE:
Premium palette

OPPOSITE TOP:
Maitre D' deli food products packaging
Design: Davies Hall

OPPOSITE MIDDLE:
A collection of Sainsbury's
Taste the Difference product packaging
Design: Parker Williams

OPPOSITE BOTTOM:
Klein Caporn sauces
Design: The Partners

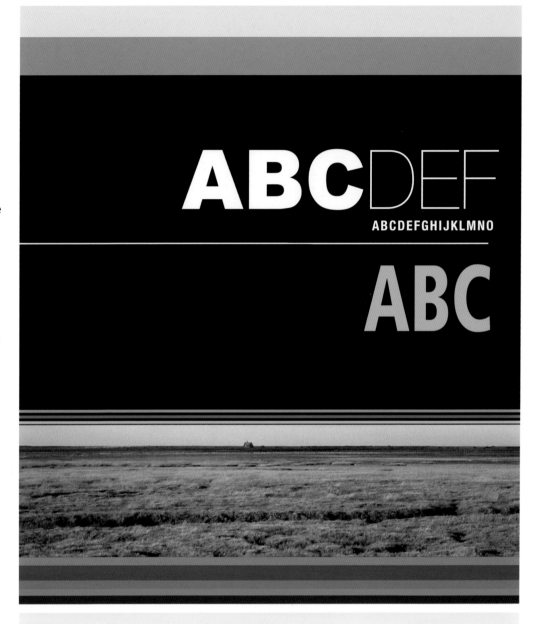

ABCDEF
ABCDEFGHIJKLMNO

ABC

ABC

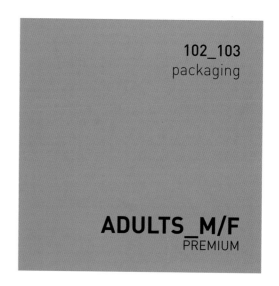

ADULTS_M/F
PREMIUM

The redefinition of luxury means that consumers have many more opportunities to experience super-premium products. Once stuffy and rare, luxury now means original, experiential, and personalized. Jewelry typifies the shift, with innovative design and handcrafted qualities counting as much as precious stones. Hence, retailers like Kabiri present themselves in a non-traditional way. With branding and packaging by Mayday, its identity is upmarket yet distinctly modern.

Vodkas have pioneered the super-premium category, capitalizing on a revitalized cocktail culture. Aloof's design for U'Luvka reflects a rich mix of ancient and modern Poland, alchemical philosophy, and Arts and Crafts references. The type identity was sympathetically refined from one of the first recorded Polish typefaces. Discretely branded, the bespoke bottle's silhouette ensures U'Luvka is instantly recognizable.

The popularity of the Mojito cocktail heralded a trend toward luxury rum brands; none more stylish than 10 Cane Rum with pack design by Werner Design Werks. Now there's an ultra-deluxe line in every spirits category. Forget preconceptions of tequila as party fuel: Cabo Uno tequila retails to discerning consumers at $225 a bottle. Designers Meat and Potatoes sourced all elements of the packaging to be made in Mexico. Housed in a handmade leather and wood box, the cork-sealed bottle comes with a crystal stopper. Each bottle of this limited edition of 21,000 bottles is numbered to earmark its allocated status as "authentic, elegant Mexican."

THIS PAGE:
Cabo Uno tequila packaging
Design: Meat and Potatoes

OPPOSITE TOP:
Innovative bottle shape for U'Luvka
Design: Aloof Design

OPPOSITE BOTTOM:
10 Cane Rum packaging design
Design: Werner Design Werks, Inc.
Kabiri jewelry packaging
Design: Mayday

ADULTS_M/F
LUXE

FOCUS _CASA LORETO

PEN PORTRAIT

Casa Loreto is a luxury Tuscan olive oil of exceptional quality, regarded in its home region as "oro liquido"—which translates to English as "liquid gold." The brand is intended for the international market, currently experiencing a huge increase in sales of super-premium single estate olive oils among sophisticated adults.

Design consultancy, The Partners was challenged to package Casa Loreto in a way that would express the preciousness of the product while creating standout within a highly saturated marketplace. Taking a lead from the oil's local epithet, the design evokes a single drop of the liquid gold escaping down the bottle toward the label, ingeniously executed in three dimensions. The pared down bottle design heroes the liquid drop, with typography kept to a minimum. All the information is expressed in Italian, emphasizing the authenticity of the product.

The packaging has won several awards, including D&AD '06 Silver for Packaging and New York Festivals '06 Finalist for Packaging. Testament to the design's originality, client, Jean Fraser-Cami enthused: "The bottle superbly expresses the desirability and quality of our oil. It certainly seems to make people very curious to taste it."

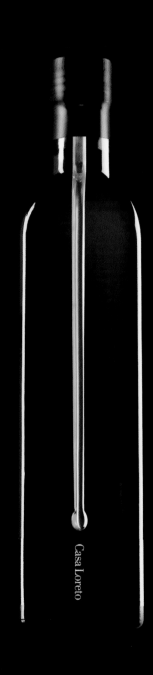

Expensive woods and leathers meet foiled type and embossed crests to create the luxe palette, which is as much about texture and finish as color and type. Referencing sixteenth-century Polish manuscripts, the pattern for U'Luvka's packaging was handdrawn by Aloof designers then printed onto flocked paper and distressed with rubbed-in wood ash. Silk-screened in a shimmering UV varnish, the pattern changes in intensity depending on the angle of light, merging with the deep burgundy background.

Signature colors play an important role, such as Veuve Clicquot's distinctive orange or the rich blue-black chosen by Identica for Russia's favorite couturier, Valentin Yudashkin.

Structure is helping to reshape luxury too. Enterprise IG helped Johnnie Walker redefine whiskey drinking rituals and thereby appeal to a younger audience of men and women with the introduction of the Johnnie Walker Gold gift pack. Using the insight that this single malt tastes best ice cold, its ingenious structure opens up to form an ice bucket. Everything about the packaging encourages sharing—moving the product away from solitary masculine usage to a convivial group activity.

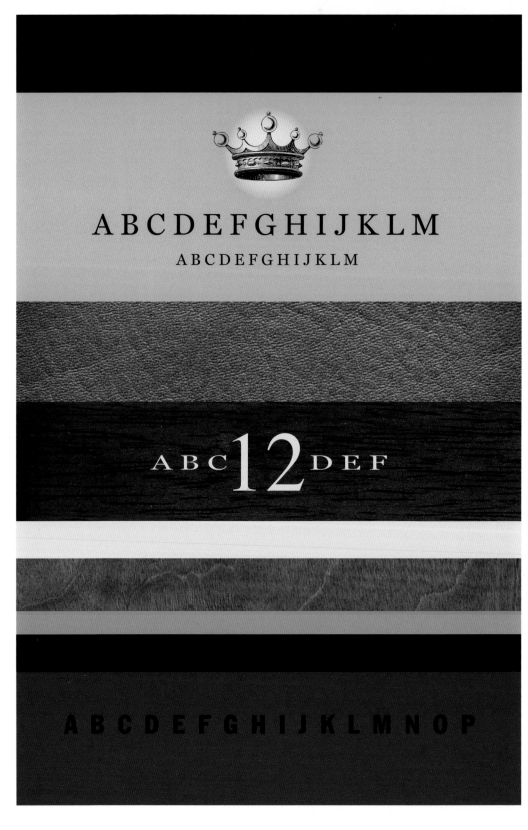

THIS PAGE:
Luxe palette

OPPOSITE TOP:
Russian designer Valentin Yudashkin's packaging
Design: Identica

OPPOSITE BOTTOM LEFT/RIGHT:
Gift packaging design for Veuve Clicquot Champagne
Design: Davies Hall
Johnnie Walker Gold Label gift box
Design: Enterprise IG
Executive Creative Director: Glenn Tutssel
U'Luvka outer box
Design: Aloof Design

ADULTS_M/F
LUXE

CASE STUDY _RDA ORGANIC

"RDA Organic is a brand that appeals to a wide audience, as it has mainstream organic appeal and allows an affordable entry point to a healthy, organic product. We wanted the packaging to reflect the transparency of the organic sector and the simplicity of the functional sector."

PATRICK AND KAREN O'FLAHERTY

FOUNDERS, RDA ORGANIC

When Karen and Patrick O'Flaherty gave up their day jobs to set up RDA Organic, they wanted to bring a fresh approach to the organic sector. They explain: "We are passionate about all things sustainable and organic and we wanted to develop healthy drinks that are convenient and delicious—in short, health shortcuts. We also got frustrated with conventional juice and smoothie companies that said they could not produce fresh organic drinks, yet still claim to be 'natural.' We believe that the only way you can ensure that products are truly natural is to be organic. Our award-winning juices are made with 100 percent organic fruit with no added sugar or artificial sweeteners, no added water, no preservatives, no flavorings, no GM, and no concentrated juices."

Having worked in marketing and finance, the pair knew the size of the opportunity. As the third largest market for organic food in Europe, UK organic sales are growing at a rate of 30 percent per year. In addition, nearly half of organic shoppers say they like to buy "distinctive organic brands." Enter RDA Organic.

The "RDA" in the name stands for "recommended daily allowance" (one bottle provides the body's full recommended daily allowance of Vitamin C).

OPPOSITE:
RDA Organic packaging
Design: Mayday

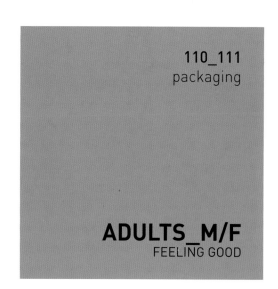

ADULTS_M/F
FEELING GOOD

"Our range is naturally functional, letting the ingredients do the work instead of added supplements," says Karen. "We worked with design consultancy, Mayday, to design a bespoke bottle and we chose labeling to reflect the transparency and goodness inside. Convenience, clear information, and a distinctive brand personality were all essential."

The O'Flahertys believe that consumers are approaching the organic sector in a different way.

RDA Organic targets not only committed mature organic consumers, but also mothers buying a healthy kids' snack, time-poor consumers wanting a convenient healthy offering, and those buying "grab and go" food and drinks. Creating a hybrid brand—part health, part convenience—reflects the broadening appeal of organic and represents a new approach to packaging healthy products.

Healthy and organic brands are shedding the horsehair shirt of worthy packaging in favor of vibrant design. Today's adults are keen to "do the right thing"—by themselves, their families, and the environment, and the feel-good benefits are being reflected in packaging with bags of personality. Using pharmaceutical grade herbs, the Dr Stuart's tea brand has a loyal following, but its soulless packaging was a barrier to winning new customers. Pearlfisher's redesign represents Dr Stuart as the idiosyncratic expert, complete with surreal illustrations and peculiar, but relevant, descriptors. White space, black typography, and color blocking brings modernity to the brand, which now carries the strapline "Extraordinarily Good Teas," redefining Dr Stuart's as a personality-driven brand that combines eccentricity with efficacy.

As the currency of organic increases, brands look to widen their appeal. Mayday used natural photography to emphasize fresh fruit ingredients and help Clearspring get listed in two major supermarkets. White helps convey purity cues, as Mayday proved with their flavor infusion treatment for Hampstead Teas. Davies Hall's redesign of Doves Farm Organic has reinforced organic in the brand icon. Simple typography and charming illustrations are the antithesis of wholemeal worthiness.

THIS PAGE:
Dr Stuart's tea packaging
Design: Pearlfisher

OPPOSITE TOP:
Clearspring organic fruit puree deserts
Design: Mayday

OPPOSITE BOTTOM LEFT/RIGHT:
Doves Farm Organic's rebranding and packaging design
Design: Davies Hall
Brand identity and packaging design for Hampstead Tea and Coffee
Design: Mayday

Clearspring

PEAR
ORGANIC FRUIT PURÉE DESSERT

Clearspring

APPLE & BLUEBERRY
ORGANIC FRUIT PURÉE DESSERT

Clearspring

APPLE & PINEAPPLE
ORGANIC FRUIT PURÉE DESSERT

Clearspring

APPLE & APRICOT
ORGANIC FRUIT PURÉE DESSERT

ADULTS_M/F

FEELING GOOD

HAMPSTEAD
TEA & COFFEE

Organic
DARJEELING
single estate leaf tea

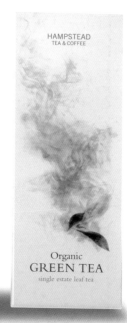

HAMPSTEAD
TEA & COFFEE

Organic
GREEN TEA
single estate leaf tea

HAMPSTEAD
TEA & COFFEE

Organic Fairtrade
FENNEL LIQUORICE
Balancing infusion

HAMPSTEAD
TEA & COFFEE

Organic Fairtrade
ROSEHIP HIBISCUS
Revitalising infusion

HAMPSTEAD
TEA & COFFEE

Organic
LEMON GINGER
Energising infusion

FOCUS _SAFEWAY O ORGANICS

Philippe Becker Design created an identity system that communicates the brand values of approachability, health, and quality across 150 products. At its heart is the bold graphic "O," not just standing for "Organics," but also found in expressive words such as "wow" (I didn't know organic foods could taste this great). The circle is also a basic shape found in nature: a symbol of wholeness, completeness, and the cycles of life.

The core brand color palette is based on the essential ingredients for life: earth, sun, water, light. Flavor colors are chosen to complement the brand color as well as the product color. The simple and clean Frutiger font family is used throughout for product name and descriptor.

Photography tells a visual story about the origin of the ingredients and communicates maximum appetite appeal. The photographic style supports the brand personality: real (using everyday settings), beautiful (not overly stylized), and fresh (plain and simple). The look and feel is contemporary, using props to accentuate the "Organic from the Source" statement. Human elements, such as hand shots, are used on the non-processed products to demonstrate freshness. Where photography was not an option, bold and graphic illustration reflects the quality of the product.

PEN PORTRAIT

Safeway O Organics is targeted at the typical grocery shopper, albeit one who's a little more "progressive" than average. Research showed that while consumers are increasingly concerned about health, wellness, and sustainability, they are no longer primarily politically motivated (as was the pioneering organic movement of the 1970s). Consumers care about quality and taste and are looking for new definitions of organic—vibrant, original, and open-minded, rather than yoga boot camp worthiness. The positioning sums this up: "we believe in organic foods for everyone."

The development of O Organics has played a major role in Safeway's strategy to be the health and wellness advocate for its shoppers. It represents a major step toward helping Safeway align its private label effort with its overall strategy of building its offerings around consumers.

Consumers are delighted by the new brand and demand has exceeded Safeway's ability to source organic supply. In one year the brand achieved over $150 million in sales—far beyond expectations and contributing to Safeway's stock price reaching a five-year high. The *Washington Post* praised the brand for its "gorgeously restrained" packaging design and in 2006 Safeway won *Brand Packaging Magazine*'s Brand Innovator Award for O Organics.

The feeling good palette reflects several trends in the health and organic sectors. The move toward a contemporary communication style is reflected in the use of more white space, bold typography, and fresh bright colors. Where once beige and earthy tones indicated goodness, now intense color conveys the uplifting benefits of healthy products, as seen in Mayday's design for Hampstead Teas' ice teas.

The shift is not simply a stylistic one. New categories are emerging that require a visual language of their won—notably, the functional and super-food categories. ® Design's approach to the packaging for Food Doctor looks natural, authoritative, and healthy, reflecting the credibility of the savvy nutritionists behind the brand. The natural benefits of hemp seed are appetizingly communicated through bold yet friendly packaging for Braham and Murray's Good range.

THIS PAGE:
Feeling good palette

OPPOSITE TOP:
Braham & Murray's Good range

OPPOSITE MIDDLE:
Packaging design for Hampstead Tea
and Coffee's Ice Tea range
Design: Mayday

OPPOSITE BOTTOM:
The Food Doctor packaging range
Design: ® Design

ABCDEFGH
IJKL
MNOPQR
STUVWXYZ

ABC
DEF

abcdefghijklmn

opqrstuvwxyz

ADULTS_M/F
FEELING GOOD

CASE STUDY _MICHAEL AUSTIN WINES

"Each wine label recounts a story loosely inspired by anecdotes of the two founders' lives...very loosely, that is."

GABY BRINK

CREATIVE DIRECTOR, TEMPLIN BRINK

The Michael Austin Winery was started in California by two long-time friends. The person Michael Austin does not actually exist—Michael and Austin are the middle names of the two founders. Approached to create the branding and packaging for the winery, design consultancy Templin Brink turned this fact into a unique angle that would help enrich every aspect of the brand's communication, inventing a fictitious character and creating stories about him to give personality and humor to each of the wines.

The stories are based on real anecdotes from the founders' lives. Bad Habit refers to how the two met as young men in a Catholic high school; High Flyer relates to the fact that one of them flies small airplanes as a hobby. Complicit in this playful fiction, the savvy customer is entertained by these far-fetched tales. Thus, we are led to believe: "Michael Austin grew up in a monastery on the outskirts of France. He was raised by a pack of wild nuns who taught him how to live on a strict diet of wine, cheese, and real estate investments. Today, he is religious about only one thing—making great wine."

OPPOSITE:
Michael Austin Wines
Design: Templin Brink

Humorous illustrations adorn the labels in a palette of deliciously muted tones. In a sector that often relies on quirky names and punchy copy to create distinctiveness, this packaging achieves the perfect balance of humor and sophistication—leading AIGA jurors to commend the design's whimsical naming and provocative style, commenting: "Everything about it has that handcrafted quality, personal touch. It feels like they made that batch just for you."

When Turner Duckworth was approached to evolve the identity of Amazon, the world's largest internet retailer, they were inspired by the energetic and happy-go-lucky approach of its founder and CEO, Jeff Bezos. Turning the generic orange slash into a smile tapped into a consumer desire to feel good about the purchases they're making and the brands they choose, as well as offering a distinctive graphic device to emphasize the A-Z inclusiveness of Amazon's offer.

Uplifting brands use design to make their consumers smile, creating a more intimate dialogue and fostering greater brand loyalty. For every bottle sold of Heavenly Wine, five percent of the price is donated to the charity Water Aid. Turner Duckworth encapsulated this "wine into water" story in the "Drink Generously" branding—a witty message that avoids appearing too worthy. Outer casings continue the humor with instructions like "Avoid feeling... FRAGILE...drink responsibly."

Target's Archer Farms is a food brand with a friendly, yet wholesome, character. Templin Brink created colorful packaging using an abstracted leaf, which became the cornerstone of the Archer Farms branding system as patterns and die-cut windows were created from it. Identica used research to arrive at a "feelgood fruit" positioning for Ocean Spray dried cranberries. The packs use quirky handdrawn type to indicate key benefits in a style that's as appealing to kids as it is to moms.

THIS PAGE:
Amazon.com packaging
Design: Turner Duckworth

OPPOSITE TOP LEFT:
Archer Farms milk packaging for Target
Design: Templin Brink

OPPOSITE TOP RIGHT,
OPPOSITE MIDDLE:
Heavenly Wine
Design: Turner Duckworth
Creative Directors: David Turner,
Bruce Duckworth
Designer: Sam Lachlan

OPPOSITE MIDDLE:
Archer Farms pasta sauce for Target
Design: Templin Brink

OPPOSITE BOTTOM:
Ocean Spray dried cranberries packs
Design: Identica

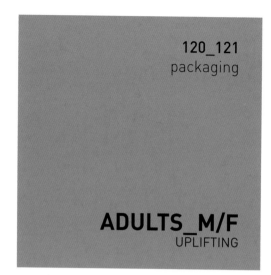

ADULTS_M/F
UPLIFTING

The uplifting palette employs warm and friendly colors with a lighthearted approach to typography and messaging. jkr's method of creating "brand charisma" is exemplified by its packaging for the London Tea Co. Using a strong modern marque as a table, chair, or sofa for a cast of characters to lounge on, the packaging deliberately feels at home in an urban context with bright colors and graphics reflecting the uplifting nature of the product inside.

Handdrawn elements convey quirkiness and approachability. Eclectic iconography can help lighten the mood of packaging while indicating features and benefits. The tone of voice is key. A neatly executed concept is The Partners' packaging for Stanley Honey—where the consumer is encouraged to plant a flower in the pot after use, in order to "keep our bees busy."

THIS PAGE:
Uplifting palette

OPPOSITE TOP:
The London Tea Co.
Design: jkr

OPPOSITE MIDDLE:
Grape Tamer and Bad Habit labels
for Michael Austin Wines
Design: Templin Brink

OPPOSITE BOTTOM LEFT/RIGHT:
Archer Farms pasta packaging for Target
Design: Templin Brink
Stanley Honey reusable pot
Design: The Partners

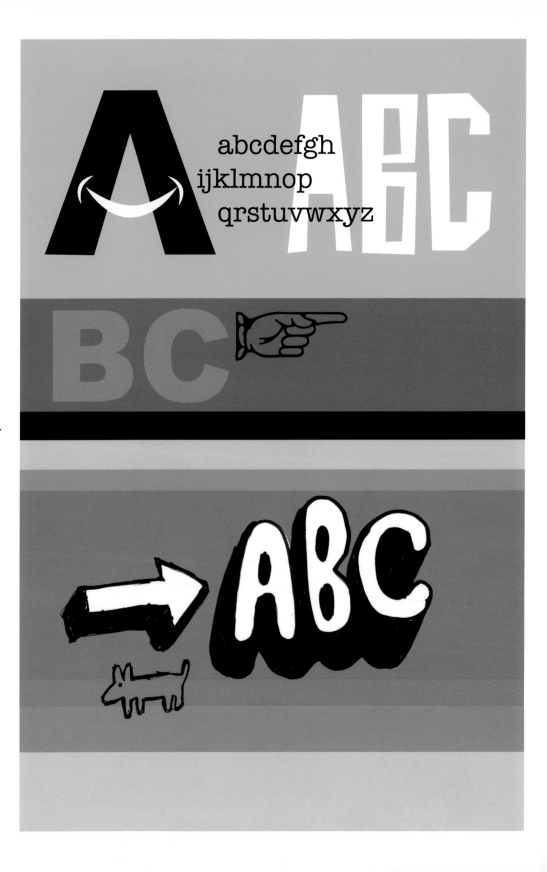

The London Tea Company

Raspberry Chilli
Gourmet Tea
Raspberry & Sweet Chilli
25 BAGS 100% ORGANIC

The London Tea Company

Lemon Retreat
Gourmet Tea
Green Tea, Lemon & Ginger
25 BAGS 100% ORGANIC

The London Tea Company

Lavender Camomile
Gourmet Tea
Lavender & Camomile
25 BAGS 100% ORGANIC

The London Tea Company

Honeybush Vanilla
Youthful Caffeine Free
100% Organic
25 BAGS

ADULTS_M/F
UPLIFTING

CHAPTER 8 _ADULTS _M

Marketing to men is big business—fashion, grooming, and accessories brands are all capitalizing on men's increasing brand savviness. While the rise of the metrosexual male blurs the distinction between products for men and women, there are still some areas that continue to be male dominated. Marketers frequently adopt a no-nonsense approach to communication for men and this extends to packaging too. Whether it's a moisturizer, a beer, or a gadget, men find reassurance in a straightforward, feature-and-benefits driven style. Most technology continues to be targeted at men, and the clarity and precision that is used to communicate these products can be transferred to other sectors, such as alcohol, dietary supplements, and even tailoring.

CASE STUDY_SKINtools

"Each new product launch moves the men's grooming category forward, but men are still more cautious than women when it comes to cosmetic products—so the male market tends to be less dynamic than its female counterpart, who if anything can be bordering on obsessive."

JONATHAN FORD

CREATIVE PARTNER, PEARLFISHER

Jonathan Ford is Designer and Creative Partner at design consultancy, Pearlfisher. He has observed the steady but sure growth of the male grooming market over several years and brought future focused consumer insight to upmarket multiple retailer, Waitrose, in a think piece that formed the brief for the SKINtools range.

Understanding and anticipating what the consumer is likely to need is at the heart of Pearlfisher's work. They do their own investigations into the changing desires that govern consumer trends. Jonathan identified several key motivating factors behind men's consumer habits, including individuality, tailored solutions, increased affluence, trust and belonging, the age blur, and the need for professional guidance and information. "Men have become multimodal," he explains. "Their needs are driven by a combination of mood, circumstance, and occasion. Duality lies at the heart of male consumerism: the desire for technological advancement, tempered by a tendency toward traditional simplicity."

Overlaying this information with a clear understanding of the Waitrose shopper further shaped the brief. "He's a mature 25-plus consumer, confident, but would still frown on the idea of a tinted moisturiser at this point."

OPPOSITE:
SKINtools
Design: Pearlfisher

The positioning for the brand is Easy Masculinity; the SKINtools name reflects trust and simplicity. "In creating such a name, it would have been easy to go down a design route of power tools and tool kits, but that would not speak to the sophisticated Waitrose shopper."

Instead, the "intriguingly simple" design employs simple and boldly elegant typography to convey dosage and directions with clarity and style.

The use of dark brown glass gives a professional, almost pharmaceutical sense of reassurance, while matte paper and foil stamping balance down-to-earth values with trust and expertise.

SKINtools
AFTERSHAVE BALM

DOSAGE:	a marble-sized amount on dry, clean skin
DIRECTIONS:	smooth into your just-shaved chin
EFFECT:	leaves skin moisturised, cooled and free from irritation

FOR MEN

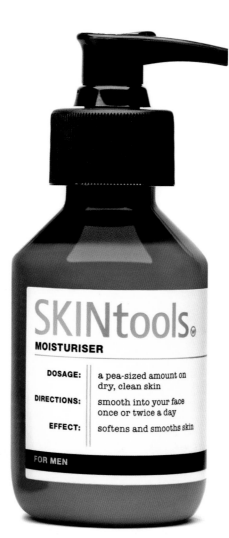

SKINtools
MOISTURISER

DOSAGE:	a pea-sized amount on dry, clean skin
DIRECTIONS:	smooth into your face once or twice a day
EFFECT:	softens and smooths skin

FOR MEN

SKINtools
FACE WASH

DOSAGE:	a marble-sized amount on already wet skin
DIRECTIONS:	massage in, mind your eyes, rinse with cool water
EFFECT:	cleansed and soothed skin

FOR MEN

It seems to be a universally accepted fact that men take charge of the barbecue. When Turner Duckworth was commissioned to redesign the barbecue range for Homebase, they decided to create some impact. Out went brown card outers, devoid of information graphics, replaced by a coherent range unified by strong black coding and photographs with visual associations: hot chillis for firelighters, graphic mustard and ketchup squiggles for disposable barbecues.

The same agency helped Levi's Dockers reposition to sell on performance, function, and technology. The branding and packaging they created works as hard as the garments, directly communicating functional benefits such as stain resistance. Men are more likely to overcome their cynicism if a product appears to have clear benefits. So Turner Duckworth's subtly masculine Naturally Active range for Liz Earle has clearly stated performance cues: "clean and fresh," "close and smooth," "calms and soothes."

The functional approach can also help to differentiate a product. Hornall Anderson Design Works created packaging for a signature beer by Widmer Brothers Brewery. Referencing the roots of craft beer brewing, the design has the spontaneous feel of a homegrown brewer's simplistic ink-stamped label, resulting in a system that resonates with the target and can easily accommodate each year's new flavors.

THIS PAGE:
Homebase Stores own brand BBQ packaging
Design: Turner Duckworth
Creative Directors: David Turner,
Bruce Duckworth
Designer: Christian Eager
Photographer: Andy Grimshaw
Retouching: Matt Kay

OPPOSITE TOP,
OPPOSITE MIDDLE:
Dockers stain defender tag
Design: Turner Duckworth

OPPOSITE BOTTOM LEFT/RIGHT:
Widmer Brothers Brewery W'05 IPA
Signature Blend packaging
Design: Hornall Anderson Design Works
Art Directors: Jack Anderson, Bruce Stigler
Designers: Bruce Stigler, Larry Anderson,
Jay Hilburn, Elmer de la Cruz
Copywriter: Amy Bosch

Liz Earle Naturally Active Skincare for men
Design: Turner Duckworth
Creative Directors: David Turner,
Bruce Duckworth
Designer: Bruce Duckworth
Account Manager: Alex Bennett

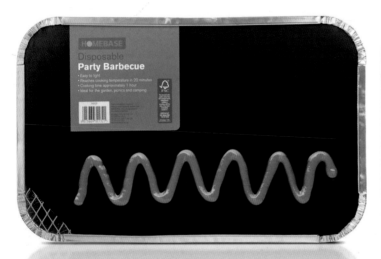

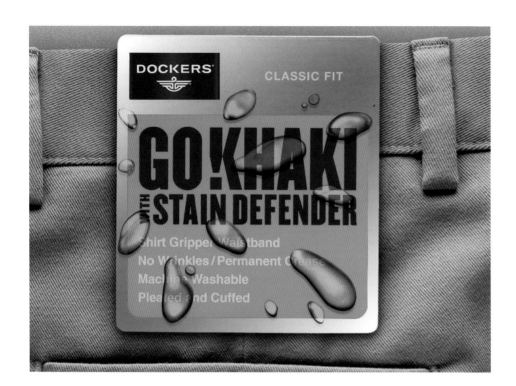

ADULTS_M
STRAIGHT TALKING

Silkwashed
Treated for an incredibly soft hand

FEEL THIS

FOCUS _BUDDY RHODES

Philippe Becker Design created a completely new category and brand from scratch, because the Buddy Rhodes products themselves had never previously been packaged and sold at retail. Buddy had a reputation, but no packages, so the design team felt it was necessary to build a brand around his name.

PEN PORTRAIT

Buddy Rhodes manufactures high-end concrete counter tops for upscale lofts, stores, and restaurants. This new line of retail concrete mixes, tints, sealers, and custom molds is targeted at contractors, interior designers, and sophisticated home decorators—predominantly 25–55-year-old men. No research was required to develop the branding, other than with Buddy Rhodes himself, who is very familiar with the target audience. They expect a high-quality product, so it had to look authentic, but not overly designed.

The overall presentation is that of a respected, serious industrial product presented in a novel way. For the logo and branding treatments, the team introduced a central "character." While not intended as a portrait, Buddy Rhodes immediately felt an affinity for the coarse, wood-cut image of a toiling character that would become the emblem of the brand. Bold type reinforces the sturdiness of the product, conveying both "retro" and "designer" aspirations.

The graphics were designed to be easily reproducible and look good using the lowest quality flexo printing on low-grade kraft paper. The "distressed" look of the printing was intentional, and the imperfections make it look like part of the system, especially on the brown paper sack. Customer feedback has been very positive and brand and product sales have grown significantly.

The straight talking palette uses subtle rather than overtly masculine design cues. Products announce their benefits by saying "this works, here's how" in a range of impactful and blocky type styles. Often, information that would usually be reserved for the back of the pack is integrated into the main pack design to reinforce notions of functional benefits.

Brown paper and earthy tones suggest no-nonsense honesty. This is used effectively by Hornall Anderson in its packaging for Leatherman Tool Group. The company wanted to fend off copycat brands with a proprietary brand image to maintain its leadership position. Another requirement was to make the clamshell-packaged product theft-proof while allowing the customer easy viewing. Hornall Anderson developed a graphic kit-of-parts including corporate identity, brand mark, typography, and color palette for each of the six tool categories. The flexible brand architecture incorporates features and line drawings. Since the allover plastic coating didn't sufficiently portray the anticipated look of quality, uncoated paper stock was fitted to the outside of the plastic to lend visual appeal to the tamper-proof packaging.

THIS PAGE:
Straight talking palette

OPPOSITE TOP:
Leatherman Tool Group packaging
Design: Hornall Anderson Design Works
Art Director: Jack Anderson
Designers: Jack Anderson, Lisa Cerveny,
David Bates, Alan Florsheim
Photographer: Condit Studio

OPPOSITE BOTTOM LEFT/RIGHT:
Homebase Stores own brand BBQ packaging
Design: Turner Duckworth
Creative Directors: David Turner, Bruce Duckworth
Designer: Christian Eager
Photographer: Andy Grimshaw
Retouching: Matt Kay
Dockers Stain Defender tags
Design: Turner Duckworth

ADULTS_M
STRAIGHT TALKING

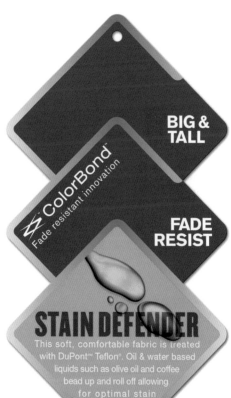

Many brands targeted at men convey a sense of high performance using science and technology references. The science of measuring time, horology, was Turner Duckworth's starting point for the creation of the Ologi brand name. The logo design incorporates the sun and the moon and the packaging invokes a time capsule, with the logo inset behind a watch glass. Turner Duckworth evokes brewing science in the packaging for Steel Reserve—a very strong premium lager at 8.1 percent alc/vol. The design uses every element to communicate its strength—from the short, dumpy bottle to the utilitarian, industrial graphics.

Geography can work too. 44° North is a premium vodka distilled from Idaho potatoes and flavored with huckleberries, a fruit indigenous to the region. To clearly differentiate the brand from the sea of "me-too" vodkas, Wallace Church sought to communicate the heart of the region in which it is distilled: 44° North is the latitude of the distillery's location. The brand's clean package design is distinctly modern and succinctly captures the purity of the Idaho landscape.

Worldbound is Werner Design Werks' concept, name generation, and brand development for Target's private label brand of luggage. Its core demographic is the seasoned traveler and working professional. The airline ticket visual language informs this design and lets the consumer know that this luggage can stand up to the rigors of world travel. The icons included in the hangtag are used on Target's entire luggage range, enabling consumers to compare the different price points and features.

THIS PAGE:
Ologi watch packaging
Design: Turner Duckworth

OPPOSITE TOP:
Steel Reserve, The Steel Brewing Company
for Mackenzie River Brewing Co.
Six pack design
Design: Turner Duckworth
Creative Directors: David Turner, Bruce Duckworth
Designer: David Turner

OPPOSITE BOTTOM LEFT/RIGHT:
Worldbound travel tag inspired swing ticket
Design: Werner Design Werks
44° North vodka bottle
Design: Wallace Church
Creative Director: Stan Church
Design Director: Lawrence Haggerty
Designer: Camilla Kristiansen

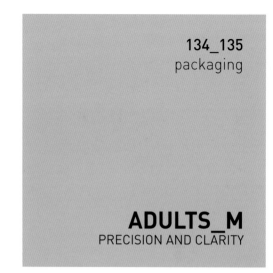

ADULTS_M
PRECISION AND CLARITY

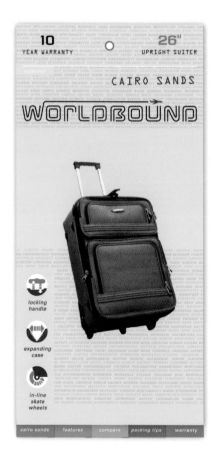

FOCUS _OZWALD BOATENG

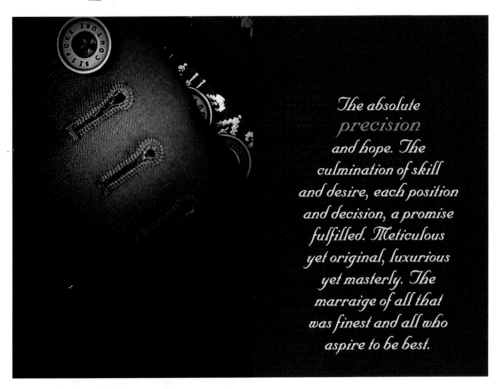

The absolute precision and hope. The culmination of skill and desire, each position and decision, a promise fulfilled. Meticulous yet original, luxurious yet masterly. The marraige of all that was finest and all who aspire to be best.

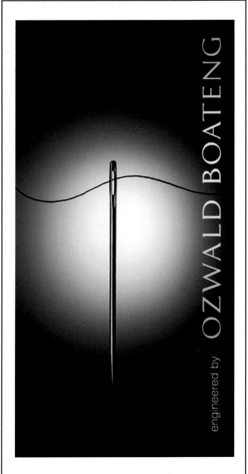

PEN PORTRAIT

Ozwald Boateng has championed the cause of British menswear and tailoring for two decades. His ability to unify classic bespoke tailoring with a unique sense of color, cut, and detail has allowed him to create a distinctive, recognizable, and widely acclaimed style that appeals to men with a sense of individuality who appreciate impeccable craftsmanship executed with a refreshingly modern approach. The brand encompasses bespoke Savile Row tailoring, a ready-to-wear collection, and a diffusion line.

Sober black and blue with silver highlights create a sense of precision and quality. A finely crafted logo with both script and sans-serif secondary typefaces supports the notion of tradition and modernity coming together. Black-and-white photography connotes clarity and luxury. Attention to detail is a key element of the brand, and this is particularly evident in the language used—"engineered by" is an original way of referring to the craft of tailoring, creating a strong sense of masculinity and stature.

The packaging structures are classic and understated. The tie pillow pack and carrier bags are manufactured using the highest quality materials in sleek black with discrete branding. The stylized brand icon is heroed while the Ozwald Boateng name appears within the folds of the pack structures.

Black, cool grays, and metallic finishes help to convey precision and clarity, highlighted with technical blues and greens. Metallic silver printed on a matte metallic teal base offers cool sophistication to Davies Hall's design for Boots SK for Men. Unfussy pack design helps focus on a product's efficacy. Russian food supplements brand, PharmaMed benefited from Identica's recommendation to simplify its brand hierarchy, focusing on the core brand and using bold, masculine packaging to bring clarity to its Men's Formula range.

A contemporary typographic treatment can help communicate a big idea. In Hornall Anderson's logo for Clearwire, the absence of the "i" reflects freedom—from wires, from hassles, from waiting. It's there—the reader completes the word in context—you just can't see it. The packaging enlists full-bleed, black-and-white photography with minimal type, so as not to distract from the overall message of delivering a simple solution to internet connections.

THIS PAGE:
Precision and clarity palette

OPPOSITE TOP LEFT/RIGHT:
Steel Reserve, The Steel Brewing Company
for Mackenzie River Brewing Co.
Design: Turner Duckworth
Creative Directors: David Turner, Bruce Duckworth
Designer: David Turner
Boots SK for Men branding and pack design
Design: Davies Hall

OPPOSITE MIDDLE:
PharmaMed rebrand and packaging range
Design: Identica

OPPOSITE BOTTOM LEFT/RIGHT:
Dockers Golf identity and packaging
Design: Turner Duckworth
Clearwire packaging
Design: Hornall Anderson Design Works
Art Director: Jack Anderson
Designers: Jack Anderson, John Anicker,
Leo Raymundo, Sonja Max, Andrew Wicklund

8.1% **HIGH GRAVITY** LAGER

THE STEEL BREWING COMPANY
LONGVIEW, TX

STEEL RESERVE®

SLOW BREWED FOR A MINIMUM OF:

28 DAYS

211

1 PINT

16 FL. OZ.

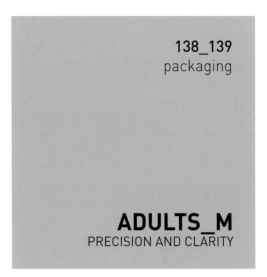

PharmaMed®

Man´s formula™

Больше
чем поливитамины™

Для обогащения организма питательными
веществами и укрепления мужского здоровья

60 капсул по 1,0 г

PharmaMed®

Man´s formula™

СпермАктин®

Для повышения количества и подвижности
сперматозоидов

30 пакетиков по 5 граммов

PharmaMed®

Man´s formula™

Антистресс™

Для повышения устойчивости мужского организма
к психоэмоциональным нагрузкам и стрессам

60 капсул по 695 мг

DOCKERS' GOLF
TOUR PANT

clearw're

clearw're

Complete broadband
Internet service in a box.

CHAPTER 9 _ADULTS _F

As women's lives embrace higher career expectations, shifting gender roles, and greater independence, products and brands aimed at them are adapting to remain relevant. According to Datamonitor, the four key drivers that women expect from products are: indulgence, convenience, performance, and sociability. These themes are all pertinent to the packaging examples seen here—from spa treatments and chocolates that signify time out from a hectic schedule to domestic products that make everyday tasks easier and more enjoyable. The have-it-all generation refuses to compromise, so as women move into motherhood, they still demand the same levels of innovation, quality, and experience they have been used to throughout their independent days.

CASE STUDY _MARIEBELLE

"I want my brand to be elegant and beautiful to the eye as well as delicious and wonderful to taste. After all, chocolate should be a complete experience."

MARIBEL LIEBERMAN

MARIEBELLE

MarieBelle is a luxury confectionery brand dedicated to creating delicious and visually stunning confections and chocolates. Established in 2001 by Maribel Lieberman, the company has grown from a small store in New York's fashionable SoHo district to an international brand that is distributed all over the world. The secret of its success surely lies in the combination of two of the target consumer's greatest indulgences: fashion and chocolate.

"Our target audience is today's gourmand," explains Maribel, "somebody who appreciates finer foods and ingredients. This translates into mostly female shoppers, aged 18–55, who read magazines and stay informed about style and trends."

Maribel is clearly not one to be constrained by the formulaic marketing approach of larger confectionery brands. With a background in fashion design, she directs all the packaging design herself—eschewing consumer research in favor of instinct, creative flair, and the confidence that she's developing packaging for consumers just like herself. The result, she points out, is a palette of colors "that are appealing from a fashion standpoint— after all, blue is typically a color that research has shown to be at odds with food brands and food marketing." →

OPPOSITE:
MarieBelle Aztec Duo Gift Set packaging
Art Director and design: Maribel Lieberman

New York
MARIEBELLE™
Aztec Hot Chocolate

AZTEC

MARIEBELLE™

→ The signature fashion-led brand style in its palette of blue and brown feels as much haute couture as maître chocolatier. Particular attention has been paid to how the packaging would look grouped together on a shelf, rather than as individual items.

Geometric squares and rectangles reflect the brand's devotion to neat, streamlined looks. Materials are always high-end—custom-dyed heavy paper, durable tins and other luxury materials and trims—because the packaging is intended for display on a desk or in a pantry, long after the confection has been consumed.

THIS PAGE:
MarieBelle's chocolate filled vanity case
Design: Maribel Lieberman

OPPOSITE:
MarieBelle's signature
blue and brown packaging
Design: Maribel Lieberman

Women appreciate everyday products that offer them a reward as an antidote to the pressures of daily life. This little bit of indulgence can come in various forms—for instance, bringing spa luxury into your own bathroom. Cowshed's primary audience is an inner-directed, luxury-seeking consumer who visits the Soho House Group of hotels and Cowshed spas at Babington House and Holland Park. With the products also available in select stores, an opportunity arose to reach a wider audience. Inspired by the interiors of Babington House, Pearlfisher's beautiful designs use wallpaper patterns of differing colors and textures to echo the benefit and ingredient of each product. A touch of warmth is added through the understated, mischievous copy, which demonstrates a unique sense of humor.

The desire to find indulgent products to meet everyday needs of gifting and self-reward was the driving force behind the launch of the Chocochic brand. In research, customers said they were frustrated with the lack of quality confectionery in beautiful packaging available for purchase in supermarkets and department stores.

Winterbotham Darby & Co briefed PureEquator to create an opulent, elegant confectionery brand, intended as more of a fashion statement than a heritage brand. Its packaging is designed to change seasonally with the introduction of different colors.

THIS PAGE:
Cowshed packaging
Design: Pearlfisher

OPPOSITE TOP:
Chocochic Easter packaging
Design: PureEquator

OPPOSITE MIDDLE:
Dianne Brill's Lash Lingerie extensions packs
Design: Dianne Brill

OPPOSITE BOTTOM LEFT/RIGHT:
MarieBelle Cacaotelle Gift Set packaging
Design: Maribel Lieberman
MarieBelle Trunk
Design: Maribel Lieberman

COWSHED | GRUBBY COW
CLEANSING MILK
made to clean by Babington House
with lemongrass and geranium
essential oils

ADULTS_F
INDULGENCE

CASE STUDY _POUT

"The Pout customer is extremely fashion conscious and not particularly brand loyal. She's always seeking new, cool, fashionable products. When she stands at the mirror in a nightclub she wants to take out a lipstick that she's proud to show off—something that says she's trendy and in the know."

CHANTAL LAREN

COFOUNDER AND CREATIVE DIRECTOR, POUT

There can be no better source of consumer insight than being your own brand's target audience. Chantal Laren, Creative Director and one of the three female founders of cosmetics brand, Pout, recalls the drive behind the brand's inception: "We wanted to create a color cosmetics environment that was feminine, aspirational, and communal—for women like us." Pout's reputation for constantly innovative packaging is the result of Chantal's infectious passion for everything that visually represents the brand.

Initially focusing on a lip range, Pout's packaging used a signature fishnet pattern and flirty, risqué product names. It quickly gained a reputation as "the underwear of makeup" among its core audience of urban professional women. Chantal reflects: "there's a close connection between lingerie and makeup. Both are empowering for a woman."

Pout has proved that packaging design can create brand loyalty and influence sales. Twenty percent of sales are driven by Pout Plump—a lip plumper packaged in a novel inflatable pack. Consumer research has informed the brand's steady evolution.

OPPOSITE:
Pout face collection
Creative Director: Chantal Laren, Pout
Design: DPAssocies, Absolute Zero°

Focus groups helped define Pout's "five minutes to fabulous" mantra. And, rewardingly, says Chantal, "consumers tell us they don't like to throw away our packaging, validating our conviction that this is a strong differentiator for the brand."

As the brand grew up with its founders, Chantal wanted to experiment with varied packaging styles to create range differential and more customer interaction. "It was important to me to keep the packaging fresh and eclectic. I considered how a fashion designer would approach a new collection, and started to think about different fabrics and prints. So for the face collection I had the idea of white stockings on skin—something quieter than the color range."
The soft peach color works with a delicate pattern inspired by tea doilies in paler tones or contrasting metallics. The design offers a coherent visual style to unify the face collection, with the flexibility to give each product an individual character.

FOCUS _POUT

PEN PORTRAIT

Aged 25–34, the Pout customer is a professional, career-minded, and highly ambitious woman. Pout knows her as an urban woman who lives life in the fast lane. She likes to create her own style. It's important to her to feel "in the know" about the latest trends and fashions. She loves surrounding herself with beautiful things. The desire to try out the latest products means that this customer is not particularly loyal, so brands need to constantly update and refresh themselves in order to retain her attention.

Pout Plump is one of the most successful ranges, with a number of applications from lashes to bust. Black and pink packaging featuring a unique lace pattern unifies products in this range. The hero product is a lip plumping gloss, which won huge customer loyalty, partly due to its innovative inflated transparent pack.

The Pout range is segmented according to application and consists of lips, eyes, cheeks, the face collection, and the Plump range. Each one has a different color and its own bespoke decorative pattern to help differentiate products in store and create more of a rich story to engage Pout's customers. The different patterns and colors of each range are chosen to complement each other in much the same way a fashion collection is put together.

Attention to detail is one of the signature elements of Pout's packaging, offering a real consumer benefit and differentiating Pout from its competitors. Pout ensures that its delicate patterns are reproduced on many different materials, from brushed-metal cases to presentation boxes, with ribbons and bows used to accessorize products. This all helps to deliver the founders' original vision of an involving, interactive brand.

If women are looking to escape life's daily grind, then the opportunity to be sexy and outrageous is the ultimate indulgence. Spoylt is an upmarket lingerie brand positioned as tempting, seductive, and impossible to resist. Pearlfisher came up with a logo concept that women see as a heart and men see as women's legs—a provocative image for both. The identity and luxurious metallic packaging makes Spoylt appropriate as a gift (from my lover) or a treat (because I deserve it).

Andy Warhol once said, "The best party I've ever been to was Dianne Brill's birthday party!" Author, model, muse, actress, TV commentator, and New York's social party whirlwind, Dianne Brill has launched a range of cosmetics with her signature style.

Featuring netting, lace, and bows and with names like Pearl Blush, Panties in a Bunch, and Lash Lingerie, the range is, like its founder, the ultimate "It Girl" fantasy.

Even if "me time" is as simple as a bath with a good book, spa treatment (if not star treatment) is available in the form of Tesco's Spa Body Therapy range, with packaging by ® Design. Intended to have broad appeal and to look good on show in the home, the range uses bold color and illustration to help segment the different offerings from serenity to vitality.

THIS PAGE:
Barbara Brudenell Bruce's
Spoylt identity and lingerie packaging
Design: Pearlfisher

OPPOSITE TOP:
Spa Body Therapy range for Tesco
Design: ® Design

OPPOSITE MIDDLE:
Dianne Brill's Pearl Blush compact
Design: Dianne Brill

OPPOSITE BOTTOM LEFT/RIGHT:
Dianne Brill's lipstick packaging
Design: Dianne Brill
Pout's Satisfy me now gift pack
Art Direction: Chantal Laren
Design: Absolute Zero°
Stella McCartney ribboned packaging
Design: The Partners

BODY Therapy SPA

SERENITY
Matahari Shower Wash

Cleanse and revive the body with
tropical monoi oil, coconut oil extract
and mimosa

BODY Therapy SPA

VITALITY
Cooling Foot Lotion

Revitalise and energise body and spir
with cooling eucalytus, menthol and
peppermint water.

BODY Therapy SPA

DETOX
Body Light Gel

Revitalise heavy legs and aid
sluggish circulation with
antioxidant seabuckthorn oil,
vitamin E, arnica, algae and
horsechestnut extracts.

BODY Therapy SPA

ANTI CELLULITE
Svelte Kit

Improve skin texture and reduce
appearance of cellulite with active
naturals seaweed, guarana and
menthol

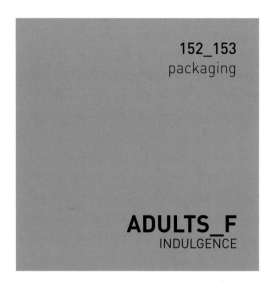

DIANNE BRILL
PEARL BLUSH

FOCUS _STELLA McCARTNEY

The Partners created a brand identity for Stella McCartney that combines precision with indulgence. The contemporary sans-serif face is rendered in a style evocative of gem-studded fabric—a glamorous brand identity, which is offered a sharper edge by the use of laser-cutting technology.

Traditional meets modern in the Stella McCartney bags, which use a combination of contrasting materials to highlight Stella's love for things old and new. Vintage ribbons were sourced to offset the precision of the laser-cut logo. The traditional linen stone paper outer contrasts with a slick, high-gloss plum interior.

PEN PORTRAIT

The clue to the identity of the typical Stella McCartney customer lies in the locations of her three flagship stores. The first boutique opened in Manhattan's achingly fashionable Meatpacking district, followed by stores in Mayfair in London and LA's West Hollywood. The denizens of these areas are extremely wealthy women, style stalwarts spurred no doubt by the kudos of counting themselves among Stella McCartney's A-list celebrity friends and clients (she designed Madonna's wedding dress).

All shades of pink are used to communicate indulgence, from the boudoir seduction of Pout to Chocochic's fashion-led fuchsia. Black makes a sexy contrast, while duck egg teamed with sumptuous brown creates a more restrained sense of opulence.

In a post-feminist world where women share responsibility with men, they are celebrating unadulterated femininity in the products they choose especially for themselves. Pretty lace, ironic bows, and romantic script all feature graphically to communicate this very personal indulgence.

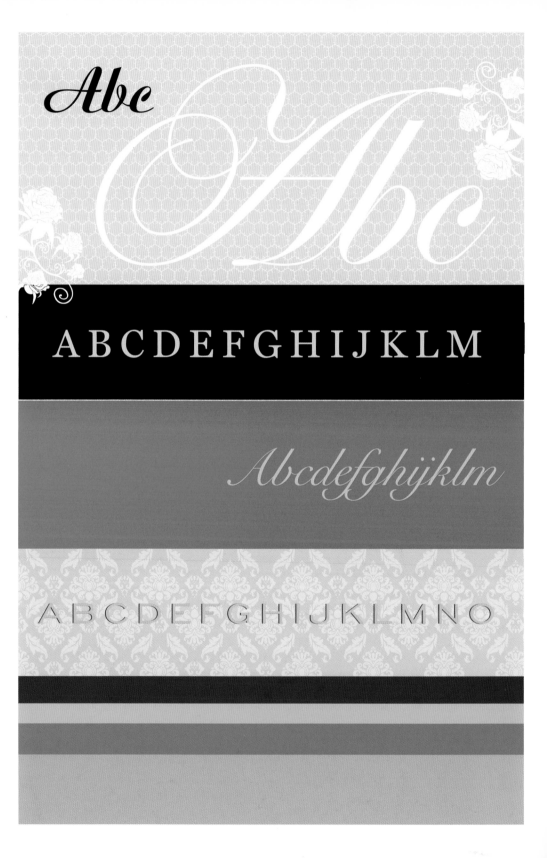

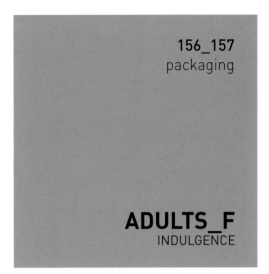

Society may be used to the high-achieving woman, but the pressure to be superwoman at home and in the office can be stressful. In response, a number of products and brands seek to lend some joy to household chores, while sharing helpful domestic hints in a world where information has ceased to be passed down from one generation to the next. Mrs. Meyer's Clean Day is a line of aromatherapeutic household cleaners. Positioned as hardworking products with the added bonus of smelling fabulous, the brand represents this work ethic with a dash of fun, through packaging by Werner Design Werks. The success of the range has resulted in a new line, created especially for pets.

Other brands offer a reassuringly nostalgic sense of homeliness, so even if you're eating a microwave meal, you can pretend the pickle on the side is homemade. Debby Bull's jams boast colorful "Really Bad" labels that are letterpress printed on pieces of vintage wallpaper by Werner Design Werks. Meanwhile, guests will think they've checked in to a boutique hotel when you leave out Wash-Sabi soap with its charmingly illustrated packaging by Pop Ink.

THIS PAGE:
Clean & Company, Mrs. Meyer's Clean Day
Carry All Cleaning Kit
Design: Werner Design Werks

OPPOSITE MAIN IMAGE:
Debby Bull's Really Bad Jam packaging
Design: Werner Design Werks

OPPOSITE MIDDLE:
Clean & Company, Mrs. Meyer's pet specific range
Design: Werner Design Werks

OPPOSITE BOTTOM:
French Paper's Pop Ink Wash-Sabi soap packaging
Design: Charles S. Anderson Design

ADULTS_F
DOMESTIC DIVA

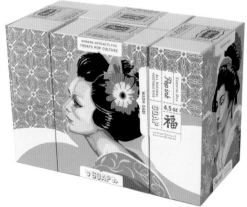

FOCUS _MOVING SENSE BY KIRSTIE ALLSOPP

PEN PORTRAIT

Most people keep their tool kits or sewing kits at the back of a drawer or hidden in an old shoebox. TV presenter and property expert, Kirstie Allsopp wanted to address this issue with a range of beautiful yet practical products. The Moving Sense range is targeted at people in the process of selling, renting, or moving into a new home, with art direction and styling connoting a female bias.

The range was developed with branding experts, Miller and Company, with design by Sebastian Conran. The packaging is a box, styled like a 1930s girls' adventure book. Original illustrations continue the theme. The products are designed to fit on to a bookshelf rather than being hidden away, so they are more likely to be at hand for impromptu little jobs. Their ingenious construction and charmingly nostalgic design makes the products attractive house-warming gifts.

The products perfectly address the current trend for making domestic chores more pleasurable and engaging. Kirstie Allsopp says of the range: "Moving house is one of life's great joys, but it often involves those niggling jobs like fixing wobbly toilet seats and dripping taps which, if left, will drive you mad within days. The tool kits help you save money by showing you how to do jobs and, unlike a lot of gifts you don't need when you move house, the kits are practical and they last."

Practical as well as stylish, the quality tools are encased in a sturdy surround. The stripy interior continues the pretty, retro theme and the "how-to" guide is styled like an old-fashioned domestic manual, complete with simple line drawings and instructive text.

STITCHES

If you've never picked up a needle and thread before, why not find some scraps of fabric and have a bit of a practice? You'll need to get to grips with four basic stitches.

Tacking Stitch

The tacking stitch (often just referred to as 'tacking' – as instructions say 'tack together' etc.) is a temporary stitch used simply to hold together two pieces of fabric. Use a single length of thread and make each stitch about 1cm long. Pin the fabrics together first, and tack, removing the pins as you go. It may seem rather long-winded, but tacking and pressing a hem first is a good way of checking that it will hang properly before you make the final repair – it's not always possible to tell how it will hang if it's only been pinned.

Fig 1: Tacking stitch

Running Stitch

Running stitching is basically the same as tacking, but uses smaller stitches – aim for 5-6mm long. It can be used for sewing together two fabric pieces, or for gathering a single piece. For example, if you want to add a frill to a cushion cover, the frill is drawn up by a row of running stitches. The more regular the stitches, the neater and more even the gathers will be. Running stitch is also a useful multipurpose stitch that can be used for many hand-sewn projects.

Backstitch

The backstitch is a very useful stitch and is the nearest you'll get to the strength and stability of a machine-sewn straight stitch. Work from left to right and you'll produce continuous stitches on the correct side of the work, and a row of overlapping stitches on the reverse. The stitches are 5mm long. Bring the thread up from the back of the fabric, make a stitch going the along through from the back of the needle back into the fabric, bringing it up from the back of the fabric, then reinsert through and bring it out further forward along the seam. The top stitches you're looking at will be smaller and along picking up the reverse side will be a continuous overlapping row of stitches then long. Use backstitching to join cushion covers, or lengths of fabric curtains etc.

Fig 2: Backstitch

Slip Stitch

The slip stitch is the most successful stitch for use on hems, because most of the thread is hidden behind the hem of the fabric, and therefore it becomes less likely that a 'pulled thread' will cause the hem to unravel. To slip stitch a hem, first fold the raw edge of the fabric twice (about 5mm is sufficient if there's little fabric available – it's just to hide the raw edges) and then pin the hem into place. Tack it if necessary. Then, with the hemmed side of the fabric facing you, slip the needle first into the folded edge and then into the single layer of fabric, picking up just a couple of 'threads' of the fabric. Bring the needle about 5mm along and back into the hem. The main stitches are hidden between the folded fabric – you're just slipping the needle in and out along the top of the turned-over hem. Remember to be careful that your stitches pick up a couple of threads on the front layer of fabric only to pick up the top of the visible from the correct side. Don't pull the stitches too tight, or else you'll get a nasty, puckered hem. It's worth spending a while perfecting your slip stitch – once you've mastered it, your items will for evermore be things of joy and beauty!

Fig 3: Slip stitch

MOVING SENSE
by Kirstie Allsopp

STARTER TOOL KIT

Another trick of the domestic diva is to make household drudgery look like a delight. Take doing the shopping. Oakville Grocery started as a country store in the heart of the Napa Valley over 100 years ago. With the addition of a new location in San Francisco, the friendly and quirky grocery destination approached Turner Duckworth to redesign its identity and packaging to reflect the store's sense of discovery. The mischievous bunny looking for tasty nibbles reminds customers that the Oakville Grocery shopping experience is nothing like the trials of a weekly supermarket haul.

Memories of his mother's home-baked delicacies encouraged restaurateur, Oliver Peyton to create Peyton and Byrne— a patisserie offering a range of foods with a reassuring sense of nostalgia. Not only is the food up to good old-fashioned standards, the packaging is beautifully considered too, courtesy of Farrow Design.

THIS PAGE AND OPPOSITE:
Oakville Grocery bag
Design: Turner Duckworth
Creative Directors: David Turner, Bruce Duckworth
Designer: Shawn Rosenberger
Illustration: Shawn Rosenberger, John Geary

OPPOSITE TOP:
Peyton and Byrne's tea and coffee range
Design: Farrow Design

OPPOSITE MIDDLE:
Oakville Grocery carrot shaped wrapper
Design: Turner Duckworth
Creative Director: David Turner
Bruce Duckworth
Designer: Shawn Rosenberger
Illustration: Shawn Rosenberger, John Geary

OPPOSITE BOTTOM:
Peyton and Byrne's preserves
Design: Farrow Design

Oakville Grocery
SINCE 1881

OAKVILLE ⚘ HEALDSBURG ⚘ SAN FRANCISCO ⚘ PALO ALTO

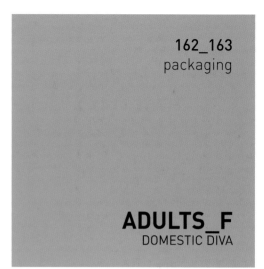

ADULTS_F
DOMESTIC DIVA

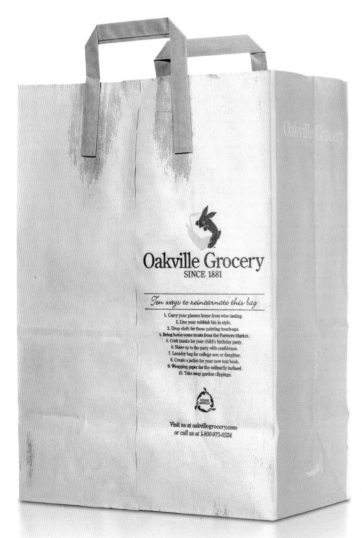

FOCUS _PEYTON AND BYRNE

Mark Farrow sourced a traditional box-making company in France that has been making boxes for 80 years. They manufactured the packaging, giving everything an authentic feel that would have been difficult to emulate with modern production methods. Far too good to hide away in carrier bags, shoppers take boxes away from the shop tied together with ribbon.

PEN PORTRAIT

Peyton and Byrne sells cakes, breads, and savories handmade each day by Roger Pizey. The brainchild of restaurateur Oliver Peyton, the concept is inspired by his mother's recipes—hence the use of her maiden name in the brand. Located between the popular homeware stores, Heal's and Habitat, Peyton and Byrne is perfectly positioned to tempt customers (largely women and metrosexual men) with the promise of post-retail gastronomic indulgence.

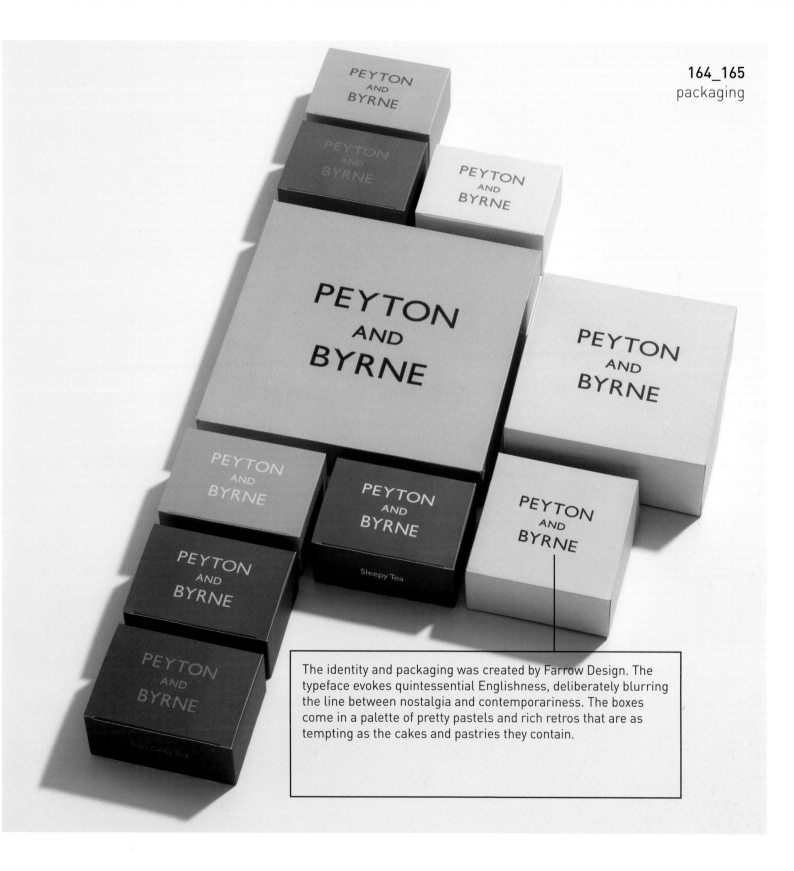

The identity and packaging was created by Farrow Design. The typeface evokes quintessential Englishness, deliberately blurring the line between nostalgia and contemporariness. The boxes come in a palette of pretty pastels and rich retros that are as tempting as the cakes and pastries they contain.

With household chores more commonly shared between men and women these days, domestic products and brands aimed specifically at women tend to communicate with a healthy dose of irony. The retro typography and 1950s-style illustration is both comfortingly nostalgic, yet knowingly naïve—its wit avoiding any accusations of gender stereotyping.

A palette of rich pastels conveys homely warmth typified by the packaging for Vivi's range of silver Christmas tree charms and cake decorations. The packaging has an artisan quality due to being printed using vegetable-based inks on 100 percent recycled stock. Designer and illustrator, Lucy Jane Batchelor created the packaging to emphasize the keepsake value of the products, while her eco-design principles were also considered appealing to Vivi's target audience.

abcdefghijklmnopqrstuvwxyz

ABCDe

abcdefghijklmnop

ABCD

ABCDEFGHIJKLM

THIS PAGE:
Domestic diva palette

OPPOSITE TOP:
Vivi's festive tree and cake decorations
Design: Lucy Jane Batchelor

OPPOSITE BOTTOM LEFT/RIGHT:
Clean Soles soap from French Paper and Pop Ink
Design: Charles S. Anderson Design
Moving Sense by Kirstie Allsopp
Developed with: Miller and Company
Design: Sebastian Conran

Christmas Charms

Designed by ViVi & illustrated by LUCY JANE BATCHELOR
www.vividesigns.com   www.lucyjanebatchelor.co.uk

ROBIN
CAKE
DECORATION
For a very splendid cake!

Designed by ViVi & illustrated by LUCY JANE BATCHELOR
www.vividesigns.com   www.lucyjanebatchelor.co.uk

ADULTS_F
DOMESTIC DIVA

MODERN ARTIFACTS FOR
TODAYS POP CULTURE

CLEAN SOLES

MODERN ARTIFACTS FOR
TODAYS POP CULTURE

CLEAN SOLES

❀ SOAP ❀

❀ SOAP ❀

MOVING SENSE
by Kirstie Allsopp

MOVING
SENSE
by Kirstie Allsopp

PICTURE
HANGING
KIT

PICTURE HANGING KIT

TOOLBOX

Yummy Mummy is a twenty-first century phenomenon. With successful career temporarily on hold, she wields a high degree of spending power, largely due to the trend toward having children later in life when couples are more financially established. Aloof Design created a brand identity, packaging, and garment labeling for fashionable maternity boutique, Elias & Grace. A two-part luggage label adorns carrier bags and gift boxes, with a baby ticket die-cut from a larger ticket, cleverly reflecting the store's mother and child clientele. Colored glass beads and metal bells strung with brown linen thread lend a playful touch to the recycled materials.

When three young mothers decided to set up Mama Mio, they knew their target market—they were it. Avoiding the worthiness of many maternity brands with the positioning "deluxe pampering for supermamas," Creative Director, Kathy Miller has designed a luxurious brand and packaging that feels every bit as indulgent as non-maternity skincare brands. Witty product names like Boob Tube keep the brand lively and fun—making it equally attractive to non-pregnant women.

THIS PAGE:
Elias & Grace identity and packaging
Design: Aloof Design

OPPOSITE:
Mama Mio's Deluxe pampering for supermamas
collection of product packaging
Creative Director: Kathy Miller

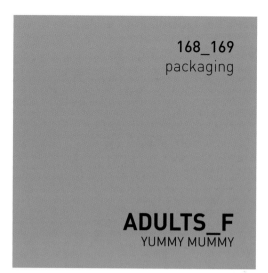

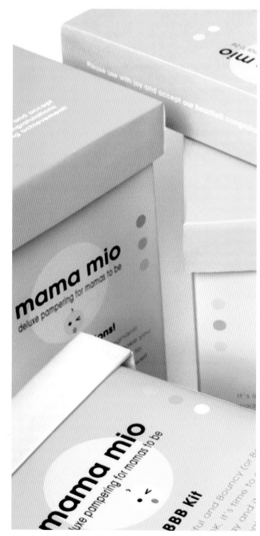

Clean, simple, and contemporary style appeals to the design conscious mother and mother-to-be. Contemporary type and modern pastels keep the tone approachable, but never twee. Warm and friendly circles feature, and even the type has a full, rounded quality. Playful language and design twists are engaging and distinctive.

The style is appealing to all thirtysomething women, with or without kids. Tend Blends' bath and body products are for stylish, professional women who have a hard time juggling the three parts of their lives: their careers, their social lives, and their personal time. Creative firm Hornall Anderson Design Works depicted the playful chic of the company and its three product lines—Work, Rest, and Play—represented as overlapping circles to reflect the diversity of women's lives. The botanic products are characterized by fun names such as MellowDrama, Creative Juices, and Nap Sugar.

THIS PAGE:
Yummy mummy palette

OPPOSITE TOP:
Tend Blends packaging range
Design: Hornall Anderson Design Works
Art Director: Lisa Cerveny
Designers: Lisa Cerveny, Holly Craven

OPPOSITE BOTTOM LEFT/RIGHT:
Mama Mio's The Smoothie Spa-at-home kit
Creative Director: Kathy Miller
Elias & Grace identity and packaging
Design: Aloof Design

ADULTS_F
YUMMY MUMMY

CHAPTER 10 _MATURE ADULTS _M/F

With maturity comes a discerning palette and, if we're lucky, the financial means to enjoy better quality products. The rise in sales of quality olive oil is indicative of the trend toward enjoying products that both taste good and offer health benefits at the same time. These consumers seek out products that offer reassuring indicators of their quality. Thus, the theme of provenance—communicating heritage and origins—is relevant as it helps to signal a product's pedigree.

CASE STUDY _RINGTONS

"Ringtons wanted contemporary packaging that at the same time embodied its heritage and tradition of knowledge and expertise. It was also very important to Ringtons that their authenticity and quality came across in the packaging."

TINA COLWELL

DESIGN DIRECTOR, DAVIES HALL

Marking its 100th anniversary, family-owned Ringtons decided to diversify from its traditional door-to-door delivery business by opening a chain of premium tea and coffee shops. The company briefed design consultancy Davies Hall to create a cohesive range of packaging across tea, coffee, food, and accessories. Design Director, Tina Colwell began by researching Rington's archive. The company had introduced different branding for each new product over the years and she was keen to find a solution that would bring consistency, while reflecting the company's authenticity.

"Ringtons are targeting a more mature consumer with some knowledge of teas," she explains, "so they didn't want to go down the fun design route, with quirky names, bright colors, and humorous illustrations." In the end, an old ration book offered inspiration for the design.

While the proposition is grown up, Tina is quick to point out that the maturity of the consumer is as much a mind-set as an age thing. "In fact, we had to try to create something relatively age and gender neutral. Some of the teas contain real flowers, but we made a conscious decision not to look too feminine."

OPPOSITE:
Ringtons Tea identity and packaging
Design: Davies Hall

With the tea divided into four "tea gardens" (fruit and herb, aromatic, classic, and rare) and coffee segmented by strength, a rich yet restrained color palette helps to differentiate each range—yellows and browns for classic teas, fragrance-inspired colors for aromatics.

A textured, uncoated stock adds to the authentic feel of the packaging. Heritage touches like the plaque-shaped labels balance the sophisticated, clean lines to create something that feels contemporary, yet pleasingly familiar.

CASE STUDY _PEDRO LUIS MARTINEZ

"The consumer in the premium wine sector is looking for solid regional cues, quality reassurance, and authenticity."

ROGER AKROYD

MAYDAY

Briefed by Direct Wines to create the packaging for a Spanish red wine, Pedro Luis Martinez, Roger Akroyd and Barry Gillibrand of branding and design consultancy Mayday worked skillfully within the category cues to create a solution that stands out from the crowd.

Mayday's approach starts with a process they term "unsettle:" an intelligent disruption of the category to understand the triggers that will create a deep response in the heart and mind of the consumer. In this case, the target was men and women in the 35–60 age range—a mature and discerning consumer that appreciates wines with authentic regional character.

"The brief was to create a classic, authentic Spanish red wine," explains Roger. "We employed striking, idiosyncratic typography and a restrained color palette to achieve essential 'established' wine credentials." The bottle is refreshingly honest and simple—deftly executed with a purposefully "undesigned" feel. The black and red type on a simple white label reflects a heritage of winemaking traditions that predates the quirky names and slick labels of some of the more showy New World kids on the block. The result exudes crafted credibility and provides all the reassurance this consumer requires.

OPPOSITE:
Pedro Luis Martinez for Direct Wines
Design: Mayday

MONASTRELL & TEMPRANILLO

BODEGAS

Pedro Luis Martínez

Jumilla

DENOMINACIÓN DE ORIGEN

2004

75cl℮15%vol

Exposicion de Vinos Nacionales
IL CONGRESO INTERNACIONAL DE LA VIÑA Y DEL VINO
EN LA EXPOSICIÓN INTERNACIONAL DE BARCELONA 1929

Diploma de El Gran Premio
Otorgado por acuerdo unánime del Jurado
Don Pedro Luis Martínez · Jumilla

PRODUCTO DE ESPAÑA
CONTAINS SULPHITES

EMBOTELLADO POR PEDRO LUIS MARTÍNEZ S.A. C/Bo IGLESIAS 55, 30.520 JUMILLA (MURCIA) ESPAÑA

Targeted at gourmets who use the highest quality ingredients, the Belazu range, designed by Turner Duckworth, communicates handpicked, handselected care through a symbol of a hand appearing from the earth as an olive tree icon. The colors evoke the feeling of the Mediterranean without overclaiming its provenance. Meanwhile, authentic Mexican chocolate displays all the qualities of its artisan provenance—simple sugar paper wrap, traditional typography, and "Since 1898" heritage reassurance.

A contemporary take on the provenance story is displayed by Enterprise IG's design for Penderyn Welsh whiskey. The label's vein of gold refers to Welsh gold mining; the wine bottle shape distinguishes it from Scotch whiskies. Mayday also uses contemporary twists to connote a product's place of origin, such as bulls' horns on Spanish olives and the use of the winemaker's name and signature, restrained ranged left typography, and a linocut illustration for Wolvehoek, a high-quality South African wine.

THIS PAGE:
The Fresh Olive Company's Belazu range
Design: Turner Duckworth

OPPOSITE TOP:
Mexican chocolate

OPPOSITE MIDDLE LEFT/RIGHT:
Penderyn single malt Welsh whiskey
Design: Enterprise IG
Executive Creative Director: Glenn Tutssel
Leisure Foods olive packaging
Design: Mayday

OPPOSITE BOTTOM LEFT/RIGHT:
Chateau Bellegrave for Direct Wines
Design: Mayday
Wolvehoek for Direct Wines
Design: Mayday

**MATURE
ADULTS_M/F**
PROVENANCE

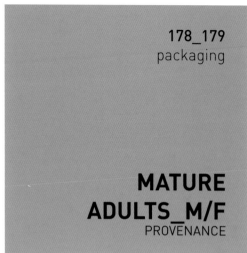

FOCUS _WAITROSE OILS

Pearlfisher's future insight research, TasteMode, showed that as consumers' knowledge of food and flavor has grown, provenance and authenticity have become key sales drivers. The three Single Estate Olive Oils in this range are from named and sourced family suppliers in three olive-growing regions of Italy—Barbarossa, Giardino di Ponente, and Podere del Sole. By focusing on the origins of these three oils, it created a range with compelling Italian credentials. The typography is designed to resemble a handmade stamp from the olive estate, and is printed on thick paper stock to add authenticity. The Waitrose name is added in silver foil—a premium stamp from a quality supermarket.

PEN PORTRAIT

Waitrose is a high-end supermarket. Famed for its food expertise and commitment to quality, it attracts consumers who are prepared to pay a little more to get the best. Waitrose consumers typically cook from scratch, have above average knowledge of food, and seek out authentic food experiences.

The design needed to reflect the exclusivity of source and "family" personality of each product. The individual character of the Ligurian, Umbrian, and Sicilian oils are brought to life using different shaped bottles and labels. Although each bottle is a standard structure, placed together they look like a bespoke family. Black glass was chosen for its premium style, but also because it protects the delicate oil from the light—a practical measure that enhances Waitrose's reputation for food knowledge.

Rich, earthy colors offer a contemporary take on heritage and help to convey a product's provenance. Signatures, stamps, and seals are all useful graphic tools that communicate quality and authenticity.

Restrained graphic treatments that don't feel overly "designed" create a sense of credibility. Templin Brink created an authentic look for Camino Real tequila using a strip of unique stamps to seal the cork. The reverse side of the label features text and graphics that can be read through the bottle from the back, while harvest imagery and the outline of the agave plant on the actual bottle form help to support the authentic "Made in Mexico" story. The same team exploited genuine Scottish heritage in the branding for MacLean wine, incorporating elements from the proprietor's family crest and patterns from the Clan MacLean tartan in the bottle, packaging, and identity system.

THIS PAGE:
Provenance palette

OPPOSITE TOP AND BOTTOM LEFT:
Camino Real Tequila
Design: Templin Brink

OPPOSITE MIDDLE LEFT/RIGHT:
MacLean Wines packaging
Design: Templin Brink

MATURE
ADULTS_M/F
PROVENANCE

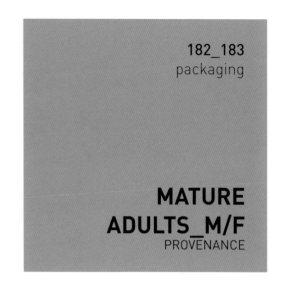

CHAPTER 11 _MATURE ADULTS _M

It's a mark of achievement and maturity for men that
they can enjoy life's rarer pleasures, whether it's a single
malt whiskey or exclusive grooming products. Two key themes
dominate products targeted at mature men. Brands with a
credible heritage offer the reassuring feeling that you're
paying for years of experience and craftsmanship.
Limited editions and numbered batches appeal to
their desire for exclusivity.

CASE STUDY _JOHNNIE WALKER

"Every element of the packaging needed to be uniquely handcrafted using the finest materials. Diageo has an archive of Johnnie Walker ephemera in Scotland, which we visited prior to the creative development. It proved to be very rich source material."

GLENN TUTSSEL

EXECUTIVE CREATIVE DIRECTOR, ENTERPRISE IG

To celebrate the 200th anniversary of Johnnie Walker's birthday, Diageo approached Enterprise IG to create a very special collectors' piece, limited to a quantity of just 200. This was not intended for sale, but to be gifted to influential people, from key industry figures to presidents.

Faced with this exciting and unique brief, Enterprise IG's Executive Creative Director, Glenn Tutssel looked to the brand's rich heritage: "Johnnie Walker used a fold-out walnut writing desk and this was the inspiration for the walnut and leather case."

The specially designed blue glass bottle was produced to replicate the finish of the nineteenth-century bottle and etched with the original quality statement. Production was strictly limited to a run of 200, with the molds smashed afterward. On the case, an angled brass plaque nods to the contemporary brand language.

Craftsmanship and attention to detail are apparent in every element of this collector's piece, which includes faithfully reproduced leather-bound books, distressed by hand. To complete the writing desk theme, Glenn recalled that the archive had displayed a book written using Johnnie Walker's original dip pen. →

→ "We sourced a supplier of similar [contemporary] ones," he continues, "but the client came back to us and said they wanted genuine antique dip pens." These were bid for at auction, so each case contains a one-off antique pen—farther increasing the value and collectible nature of the set. It is rumored that one was auctioned on the internet for $26,000.

Turning to a saleable limited-edition piece to commemorate the same anniversary— this time 4,000 items, selling for $4,000 each—the team collaborated with Baccarat to design and produce a crystal decanter that would act as a lasting, refillable piece of tableware.

The decanter is contained in a stitched blue leather case, custom-made and lined with cream chamois; its stopper is engraved with the Walker image. With the insight that some collectors purchase two—one to keep and one to share with friends—Glenn and his team ensured the piece exuded British craftsmanship for maximum appeal in duty free.

THIS PAGE AND OPPOSITE:
Johnnie Walker 200th Anniversary collector's piece
Design: Enterprise IG
Executive Creative Director: Glenn Tutssel

Rich materials like walnut and leather help to evoke a sense of heritage and tradition. Deep blues and highlights of gold feature, as well as club tie colors for a masculine sense of nostalgia, as seen on the ribbon of Penhaligon's Douro Cologne. Fountain pen signatures suggest craftsmanship and motifs such as fleur-de-lys, taken from coats of arms, farther create a sense of heritage. Issue numbers offer the idea of limited-edition batches and associated rarity, while dates create an established feel.

Penhaligon's fragrances conjure up a world of traditional gentlemen's grooming with apothecary-style bottles and outer packaging featuring Victorian engraved illustrations, suggestive of a gentleman's closet. Crest-like holding devices like the one used on Endymion create heraldic associations.

THIS PAGE:
Heritage palette

OPPOSITE:
A collection of men's fragrance packaging
Design: Penhaligon's

OPPOSITE BOTTOM RIGHT:
Johnnie Walker 1805
Design: Enterprise IG
Executive Creative Director: Glenn Tutssel

DOURO
COLOGNE

MATURE
ADULTS_M
HERITAGE

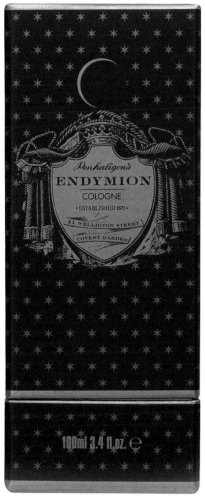

Penhaligon's
ENDYMION
COLOGNE
· ESTABLISHED 1870 ·
41 WELLIGTON STREET
COVENT GARDEN

100ml 3.4 fl.oz. e

Penhaligon's
ENDYMION
COLOGNE
· ESTABLISHED 1870 ·
41 WELLIGTON STREET
COVENT GARDEN

CHAPTER 12 _MATURE ADULTS _F

Today's mature female consumer is active, well informed, and adventurous. She has greater spending power than ever before. More likely to feel the pressures of responsibility in later life, she sees the appeal of "back to nature" escapism, though contemporary packaging combines this romanticism with the benefits of natural plant efficacy. This consumer buys into the philosophy of "feel good on the inside, look good on the outside" and wellbeing is a theme picked up by many brands that claim to offer a holistic approach to health and beauty.

Archive prints can be a source of inspiration to create a romantic theme, especially when reproduced using traditional methods. Department store Marshall Field's teamed up with the Art Institute of Chicago to release a limited-edition range of collectible gift cards, featuring masterpieces from the Institute's collection. Wink Design created the name and packaging for Field's Flora, so named because most of the artwork selected from the collection has a floral theme. Letterpress printing on specially selected paper creates a sense of nostalgia, giving the impression that this may have been something uncovered in a grandmother's attic, like a vintage flower seed packet. A limited-edition numbered soap box containing all four collectible giftcards also includes a letterpressed card, allowing the owner to personalize the box.

THIS PAGE:
Marshall Field's limited-edition
Art Cards gift card sleeves
Design: Wink Design

OPPOSITE:
Marshall Field's
Art Cards soap box
Design: Wink Design

OPPOSITE BOTTOM RIGHT:
Marshall Field's
Art Cards direct mailer
Design: Wink Design

MATURE
ADULTS_F
FLORAL

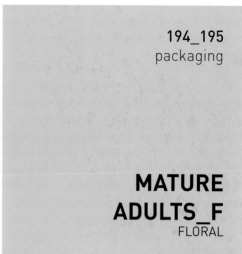

FOCUS _PENHALIGON'S

PEN PORTRAIT

Penhaligon's customers value heritage, quality, and style. The name is synonymous with classic fragrances and stores in the UK can be found in historic cities such as Windsor, Edinburgh, and York. Internationally, the brand has a presence in Paris and New York. While the typical Penhaligon's customer might be characterized as mature, the brand ensures that it remains relevant, with new store design and store openings offering a fresh, contemporary direction while retaining the traditional heritage.

Classic shapes and materials are essential to Penhaligon's style. Glass bottles with traditional stoppers refer back to the brand's history; the company was founded in 1870 by Henry Penhaligon, a barber who created a collection of personal grooming items for his wealthy customers. Many of the products carry royal warrants, supporting this credible heritage. The signature ribbon round the neck of the bottle was originally a practical way of catching any spillage of the scent from the bottle. Bluebell and Lavandula are two of the brand's most enduring fragrances. The simple approach of letting the floral scents speak for themselves is often emulated by contemporary brands, as we become more aware of aromatherapy and the natural efficacy of plants.

Flower photography and historic illustration in soft pastels create a romantic floral theme. Pinks and mauves feature alongside horticulture-inspired muted greens. Serif type creates a traditional feel, handwriting creates a personalized touch, and script faces offer a romantic flourish.

Davies Hall was commissioned to design the identity and packaging for Boots Flower Therapy. Created to appeal to a 30–55 female audience, the range consists of over 40 lines including body lotions, soaps, bath essences, candles, and linen spray in three fragrances: Rose & Chamomile, Orange Flower & Calendula, and Lavender & Elderflower. Flower photography by Carol Sharp is used to evoke the senses and communicate the efficacious nature of these products. A simple wraparound banner carries the handwritten identity.

Woolworth's Fleur Bath and Body Range features packaging by ® Design. The designers took into consideration the age of the consumer and the style of their homes, using a floral print and soft color in conjunction with classic typography and feminine embellishments like ribbon.

THIS PAGE:
Floral palette

OPPOSITE TOP AND BOTTOM RIGHT:
Boots Flower Therapy packaging
Design: Davies Hall

OPPOSITE MIDDLE RIGHT AND BOTTOM LEFT:
Fleur range for Woolworths
Design: ® Design

ABCDEFGHIJKLMNO

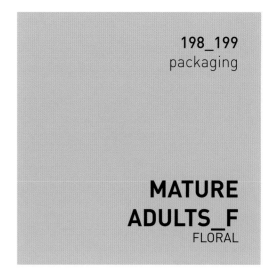

**MATURE
ADULTS_F**
FLORAL

The desire for natural wellbeing is being picked up by brands aimed at a number of consumer groups. For more mature women, the emphasis is on tranquillity and the efficacy of natural ingredients. Turner Duckworth's packaging for Liz Earle's Naturally Active Skincare range features a logo of a plant that turns into a flower made up of molecular diagrams to communicate the meeting of science and nature. The soft pastel color palette uses silver for a technical edge. For the brand's Christmas gift packaging, photography of magical forest landscapes suggests serene natural beauty.

For Superdrug's range of natural herbal supplements, Turner Duckworth looked to the herbs themselves for inspiration, Graphic, kaleidoscopic images made up of the herbs create an uplifting brand language that offers maximum differentiation between the various offerings in the range. Russian nutritional supplements brand, PharmaMed pursues a simple branding approach, courtesy of Identica, with clean stripes in colorways that differentiate products for the over-50s woman from others in the range.

THIS PAGE:
Liz Earle Christmas packaging
Design: Turner Duckworth
Creative Directors: David Turner, Bruce Duckworth
Designer: Bruce Duckworth
Account Manager: Alex Bennett

OPPOSITE TOP:
Superdrug Herbal Supplements
Design: Turner Duckworth
Creative Directors: David Turner, Bruce Duckworth
Designer: Jamie McCathie
Photographer: Carol Sharp
Retouchers: Peter Ruane, Reuben James

OPPOSITE BOTTOM LEFT/RIGHT:
Liz Earle Naturally Active Skincare range
Design: Turner Duckworth
Creative Directors: David Turner, Bruce Duckworth
Designer: Bruce Duckworth
Account Manager: Alex Bennett
Pharmamed 50+ Life Formula
Design: Identica

Milk Thistle Extract
Standardised extract of Silybum marianum

30 TABLETS
150mg

Black Cohosh Extract
Standardised extract of Cimicifuga racemosa

30 TABLETS
250mg

Ginkgo Biloba Leaf Extract
May help to maintain a healthy peripheral circulation, particularly to the hands, feet and brain.

30 TABLETS
40mg

200_201
packaging

MATURE ADULTS_F
WELLBEING

SUN SHADE™ SPF24 FACE PROTECTOR
NATURALLY ACTIVE INGREDIENTS
GREEN TEA, POMEGRANATE, AND NATURAL VITAMIN E
Non-chemical sunscreens
Fragrance-free

NATURALLY ACTIVE SUNCARE
LIZ EARLE

75ml ℮ 2.5 fl.oz.

SUN SHADE™ BOTANICAL SELF-TAN SPRAY
NATURALLY ACTIVE INGREDIENTS
NATURAL VITAMIN E, ORGANIC ALOE VERA, SWEET ORANGE AND GERANIUM
For a natural healthy glow

NATURALLY ACTIVE SUNCARE
LIZ EARLE

95ml ℮ 3.2 fl.oz.

NOURISHING BOTANICAL BODY CREAM™
NATURALLY ACTIVE INGREDIENTS
ECHINACEA, NATURAL VITAMIN E AND PURE ORANGE, ROSEWOOD AND ROSE-SCENTED GERANIUM ESSENTIAL OILS

NATURALLY ACTIVE BODYCARE
LIZ EARLE

50ml ℮ 1.6 fl.oz.

ENERGISING HIP AND THIGH GEL™
NATURALLY ACTIVE INGREDIENTS
ORGANIC DAMASK ROSE WATER, GINKGO BILOBA, HORSE CHESTNUT, BUTCHER'S BROOM, IVY AND 10 PURE ESSENTIAL OILS
Tones, firms and softens

NATURALLY ACTIVE BODYCARE
LIZ EARLE

150ml ℮ 5 fl.oz.

CLEANSE & POLISH™ HOT CLOTH CLEANSER
NATURALLY ACTIVE INGREDIENTS
ALMOND MILK, ROSEMARY, CHAMOMILE AND EUCALYPTUS ESSENTIAL OIL
Cleanses and gently exfoliates for smoother, clearer skin

NATURALLY ACTIVE SKINCARE
LIZ EARLE

100ml ℮ 3.3 fl.oz.

NATURALLY ACTIVE SKINCARE
LIZ EARLE

SKIN REPAIR MOISTURISER™ NORMAL/COMBINATION
NATURALLY ACTIVE INGREDIENTS
BORAGE AND AVOCADO OIL, ECHINACEA, BETA-CAROTENE, AND NATURAL VITAMIN E
Nourishes, smoothes and strengthens for a natural healthy glow

50ml ℮ 1.6 fl.oz.

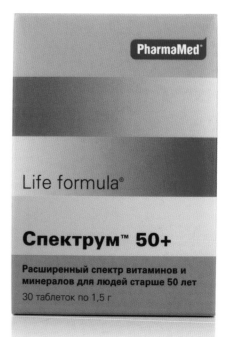

PharmaMed®

Life formula®

Спектрум™ 50+

Расширенный спектр витаминов и минералов для людей старше 50 лет
30 таблеток по 1,5 г

Calming tones of aqua and mauve tie in with natural, earthy colors to create the wellbeing palette. Type is simple sans-serif. Photography of natural elements such as pebbles is used abstractly to evoke naturally tranquil environments.

Davies Hall's packaging for Boots Healthy Living range was conceived as an identity that would be immediately recognizable as a wellbeing range without the need for a formal sub-brand. The livery had to be adaptable to work across 200 lines with products in three key areas: sleep, energize, and relax.

Turner Duckworth's solution to bring together disparate footcare products in Superdrug's range was to create a footprint motif relative to each subcategory, so bubbles connote relaxing, while pebbles suggest exfoliating. The Neal's Yard signature blue bottle suggests apothecary expertise, but was being copied by cheaper brands. To help restate the brand's authority in natural remedies, Turner Duckworth revised the brand identity, using a tree to depict plant extracts, while vibrant label colors signpost products and add a contemporary edge.

THIS PAGE:
Wellbeing palette

OPPOSITE TOP:
Boots Healthy Living
Design: Davies Hall

OPPOSITE MIDDLE RIGHT
AND BOTTOM RIGHT:
Superdrug Footcare range
Design: Turner Duckworth

OPPOSITE BOTTOM LEFT:
Neal's Yard natural remedies
Design: Turner Duckworth

abcdef
ABCDEFG

ABCDEFGH

Flex ring

Kaisoku mat

Aromatic lamp

Battery Operated
Celluline Massager

superdrug
relaxing foot soak
With lemon, tea tree & camomile
Soothes & softens

superdrug
exfoliating foot scrub
With lemon, tea tree & walnut husk
Soothes & invigorates

STIMULATING BATH OIL

CALENDULA SHAMPOO
FOR DRY SKIN

LEMONGRASS SUN LOTION

GERANIUM & ORANGE MASSAGE OIL

ELDERFLOWER CLEANSING LOTION
FOR OILY AND PROBLEM SKIN

BABY BATH

MADE IN ENGLAND

APPENDIX _FEATURED DESIGNERS

ABSOLUTE ZERO DEGREES
www.absolutezerodegrees.com

ALOOF DESIGN
www.aloofdesign.com

BOON INC
www.booninc.com

B&W STUDIO
www.bandwstudio.co.uk

CHARLES S. ANDERSON DESIGN
www.csadesign.com

DAVIES HALL
www.davieshall.co.uk

DALZIEL + POW
www.dalziel-pow.co.uk

DP ASSOCIES
www.dpassocies.com

EASY TIGER CREATIVE
www.easytigercreative.com

ENTERPRISE IG
www.enterpriseig.com

FARROW DESIGN
www.farrowdesign.com

FACTOR DESIGN
www.factordesign.com

**HORNALL ANDERSON
DESIGN WORKS**
www.hadw.com

IDENTICA
www.identica.com

JKR
www.jkr.co.uk

KIDROBOT
www.kidrobot.com

KINNERSLEY KENT DESIGN
www.kkd.co.uk

LUCY JANE BATCHELOR
www.lucyjanebatchelor.me.uk

MAYDAY
www.maydaylivingbrands.com

MEAT AND POTATOES
www.meatoes.com

PARKER WILLIAMS
www.parkerwilliams.co.uk

PEARLFISHER
www.pearlfisher.com

PHILIPPE BECKER DESIGN
www.pbdsf.com

PURE EQUATOR
www.pure-equator.com

RUNE MORTENSEN
www.runemortensen.no

SMART DESIGN
www.SmartDesignWorldWide.com

TEMPLIN BRINK
www.templinbrinkdesign.com

THE PARTNERS
www.thepartners.co.uk

TURNER DUCKWORTH
www.turnerduckworth.co.uk

UTILIA
Jiri Vanmeerbeeck
www.utilia.be

WALLACE CHURCH
www.wallacechurch.com

WERNER DESIGN WORKS INC
www.wdw.com

WINK
www.wink-mpls.com

YANG RUTHERFORD
www.yangrutherford.com

PHOTOGRAPHY CREDITS

Chinese toothpastes p38 + p41
Jon Warren

Silver Clouds photography p57
Michel Barigand

INDEX _

INDEX _

ACKNOWLEDGMENTS _

"Many thanks to everyone who contributed to this book, including the team at RotoVision— April, Jane, and Tony—and Chris Middleton for his initial concept and support.

We would like to thank Michael Peters OBE for the introduction and everyone who took the time to be interviewed for the special case studies.

Special thanks also to Vicci Baigrie, Rob Hall, Nicola Pallett, Ian Keltie, and Spike."

MARK HAMPSHIRE and KEITH STEPHENSON